The Militarization and Weaponization of Space

The Militarization and Weaponization of Space

Matthew Mowthorpe

LEXINGTON BOOKS
Lanham • Boulder • New York • Toronto • Oxford

LEXINGTON BOOKS

Published in the United States of America
by Lexington Books
An imprint of The Rowman & Littlefield Publishing Group, Inc.
4501 Forbes Boulevard, Suite 200, Lanham, Maryland 20706

PO Box 317
Oxford
OX2 9RU, UK

Copyright © 2004 by Lexington Books

All rights reserved. No part of this publication may be reproduced, stored in a retrieval system, or transmitted in any form or by any means, electronic, mechanical, photocopying, recording, or otherwise, without the prior permission of the publisher.

British Library Cataloguing in Publication Information Available

Library of Congress Cataloging-in-Publication Data

Mowthorpe, Matthew, 1973–
 The militarization and weaponization of space / Matthew Mowthorpe.
   p. cm.
 Includes bibliographical references and index.
 ISBN 0-7391-0713-5 (cloth: alk. paper)
 1. Astronautics, Military. 2. Space warfare. 3. Space weapons. I. Title.

UG1520.M69 2004
358'.8—dc22   2003060531

Printed in the United States of America

∞™ The paper used in this publication meets the minimum requirements of American National Standard for Information Sciences—Permanence of Paper for Printed Library Materials, ANSI/NISO Z39.48–1992.

Contents

Acknowledgments		vii
Introduction		1
1	The United States Approach to Military Space during the Cold War	11
2	The Origins of Ballistic Missile Defense in the United States and the Soviet Union and the ABM Treaty	35
3	The Soviet/Russian Approach to Military Space during Cold War and Beyond	55
4	China's Military Space Program	83
5	The United States and Soviet ASAT Programs	109
6	The Space-Based Laser for Ballistic Missile Defense	140
7	The Revolution in Military Affairs and the Militarization of Space	165
8	The Post-Cold War Military Space Policy of the United States	185
Conclusion		211
Appendix		221
Bibliography		227
Index		249
About the Author		253

Acknowledgments

I would like to thank Dr. Eric Grove for his comments and direction during the writing of this book. I would also like to thank Dana Johnson whose insightful comments and research material deserves a special note. I would also like to thank Dr. Paul Robinson and Dr. Bhupendra Jasani for their helpful comments. Any errors contained within the book are, however, solely the author's.

During the writing of this book I would like to thank my parents, sister, grandma, and my granddad, who sadly died while this book was being written. They each in their own ways helped enable me to bring this book to fruition. A further thanks must go to Rui Yi who helped me during the writing of this book.

Introduction

The Militarization and Weaponization of Space

Since the dawn of the space age a considerable literature has been created on the militarization and weaponization of space. In the 1980s this was subjected to a form of hijacking by the Strategic Defense Initiative.[1] The number of books and articles on the SDI program overwhelmed the more general military space literature. The SDI debate covered the possibility and practicability of various kinds of space systems designed to intercept ballistic missiles. The enthusiasm President Reagan injected into his speech announcing the Initiative appeared to make many believe that almost anything was technologically possible in orbit. Although ballistic missile defense remains a major issue, the debate has since broadened with the announcement of an "Revolution in Military Affairs" based on space surveillance and command and control platforms.[2] This has made the issue of space control central to the contemporary strategic debate. It is timely therefore to survey afresh the subject of military space putting technological possibilities into a broader strategic context.

Thinking on the military uses of space has evolved since Sputnik 1 went into orbit in 1957. The principle was established at an early stage that bodies beyond the atmosphere were not subject to normal restrictions on rights of overflight. This was extended into the "sanctuary" concept, discussed more fully in the next chapter, that sought to maintain space as a medium where surveillance and communications platforms could operate unchallenged.[3] Although the desire to maintain space as a strategic sanctuary remained strong with many political leaders the development of technology allowed more adventurous philosophies to develop. The "survivability" school emphasized the need to protect vital satellite systems.[4] The "space control" school went further and argued for the use of weapons against an enemy's space assets to deny him the use of the medium. The final school of thought, the "high ground," emerged at the time of the SDI speech, and saw space as the medium from which the earth could be dominated.[5]

The SDI literature discussed a range of systems such as X-ray lasers,[6] particle beams, and kinetic energy interceptors that were far from technological maturity. It was argued, misleadingly, that these systems were just around the corner. On the other side, critics emphasized the constraints of the Antiballistic Missile Treaty and argued for a wider arms control regime to limit the weaponization of space. The military space literature expanded to include discussion of matters of international law. The SDI speech therefore had two effects on the military space literature: it led to a burgeoning amount of technology-based work, and added an international legal perspective.

The end of the Cold War saw a shift in emphasis in the debate. The campaign in the Persian Gulf and the liberation of Kuwait redefined the military space agenda and stimulated discussion of a "Revolution in Military Affairs." The use of ballistic missiles by Iraq also focused attention on the threat of ballistic missile attack, and added impetus to a redefined missile defense program.

Because of the closed nature of the USSR, the Soviet military space literature was not as vast as the American. The available material, however, provides the basis for an account of what space assets the Soviet Union had during the Cold War period. These included reconnaissance, communications, meteorology, antisatellite, and early warning satellites. The Soviets also deployed a ballistic missile defense system which used nuclear armed missiles for both exoatmospheric and endoatmospheric interceptions. The end of the Cold War saw a reduction in Russia's military space activities but the post-Communist regime still realized the importance of military space systems and reorganized its space forces in order to keep their military space program active.

The military space literature of the People's Republic of China (PRC) is as limited as that of the Soviet Union and for similar reasons. Nevertheless, the PRC has made significant progress in developing military space capabilities. These include reconnaissance, communications, and meteorology satellites. The PRC also has three antisatellite programs which indicates a resolve to obtain such a capability. One particular aspect of the PRC's military space program is the extensive use of cooperative programs with a significant number of countries. This cooperation enables the PRC to develop capabilities and systems quicker and more efficiently than otherwise would be the case.

This thesis analyzes the military space policies of the United States, Soviet Union/Russia and the PRC from the Cold War period to the present day. It focuses on the major issues of the development of ballistic missile defense and antisatellite systems and assesses how far space will be weaponized as well as militarized. This is of special importance given the vital role played by space platforms in the Revolution in Military

Affairs (RMA). It does not specifically address the military space programs of France and NATO, although they did, and in some cases continue to have, significant military space programs. This is done since the focus of the thesis is on the weaponization of space and hence the countries that have developed and continue to develop these programs are analyzed.

The militarization of space was initially a product of the rivalry between the United States and the Soviet Union and the desire of the United States to examine what the Soviet Union was doing by developing a satellite reconnaissance program. The term militarization of space means the use of assets based in space to enhance the military effectiveness of conventional forces or the use of space assets for military purposes. The military purposes of space expanded to include communications, electronic intelligence, photoreconnaissance, meteorology, early warning, navigation, and weapons guidance. The importance of military space to both the United States and the Soviet Union grew to enormous proportions during the Cold War and many argued that the area should be kept free from weapons. As a partial step towards this situation the Outer Space Treaty of 1967 banned weapons of mass destruction from orbit and demilitarized heavenly bodies. The militarization of space is distinct from the weaponization of space, which is defined as either weapons based in space or weapons based on the ground with their intended targets being located in space. Both the United States and the Soviet Union researched and developed antisatellite weapons. The Soviet Union also developed a capability to de-orbit bombs from space, known as a fractional orbital bombardment system.

The near consensus on the nonweaponization of space saw signs of strain during the latter part of the Cold War, especially with regard to the Strategic Defense Initiative announced in the United States. Since the demise of the Soviet Union it has been put even more into question. The United States has been researching and developing space weapons with serious prospects of success. These take the form of ballistic missile defense systems incorporating an exoatmospheric interceptor designed to intercept a threat ballistic missile in space. The United States is also developing an antisatellite weapon system in two forms. One is a kinetic energy weapon system used to ram a satellite in orbit. The other uses directed energy in the form of a laser to damage the sensitive electronic components such as the solar panels and optical sensors contained on the satellite. Other space weapon concepts include kinetic energy weapons based in space with the ability to strike ground targets. The United States is not the only country interested in developing space weapons. Both Russia and the PRC are researching and developing antisatellite capabilities.

Space is becoming more important because of advances in the field of information technologies, sensors, computing, and telecommunications on the battlefield. These form the basis of the "Revolution in Military Affairs" (RMA). The RMA is reliant on military space systems. They provide much of the sensor information which is utilized when acquiring an illumination of the battlefield. Space also provides the means for the transmission of the sensor information that allows the battlefield commander to know much more than a less well-equipped opponent. If the ability to utilize these space systems is hampered then the enabling edge these assets provide will be affected detrimentally. The concept of the RMA has serious implications for the militarization and weaponization of space.

Another trend that has considerable implications for the militarization of space is the proliferation of ballistic missiles. Ballistic missile capabilities in the hands of "rogue states" provide the main current rationale for the deployment of national missile defense (NMD) by the United States. The requirement for a ballistic missile defense system implies developments in military space systems. Space-based infrared sensors could be used to track the ballistic missile or space-based systems used as the kill mechanism.

The events in the Persian Gulf in 1991 reiterated the importance of military space. The use of the Global Positioning System enabled navigation to occur in the featureless desert and enabled troops to advance rapidly in pursuit of the liberation of Kuwait. The increasing use of military space was also demonstrated with the events in Yugoslavia with the campaign against the Serbian attacks on the Kosovan Albanians. Tomahawk cruise missiles were guided by Global Position System satellites to increase their accuracy still further. The use of Joint Direct Attack Munitions that used satellite navigation systems provided a cheaper option compared with the Tomahawk cruise missile. The evolution of military space systems during recent campaigns will be analyzed in chapter eight.

The first chapter examines the United States approach to military space during the Cold War. This chapter outlines the theories of military space power that have emerged during this period, namely the sanctuary view of space power, the survivability school, the space control view, and the high ground school of thought. These views of space power are incorporated to analyze the development of military space policy by the United States during the Cold War period. The individual presidencies are analyzed in turn from the Eisenhower period up to but not including the Clinton administration. This chapter concentrates on the contention that there was a consensus formed regarding the military uses of space which sought to prevent space from being weaponized. This does not mean that the concept of space weapons was not addressed. The research and development of space weapons did take place during this period. Towards the latter part of the Cold War period with the announcement of the Strategic Defense Initiative a straining of this consensus began to occur. This chapter analyzes the space-based ballistic missile defense programs that were outlined during the Strategic Defense Initiative and its subsequent reorientation, in the form of the Global Protection Against Limited Strikes system that was announced during the first Bush administration.

The second chapter focuses on the issue of ballistic missile defense, its origins and development in both the United States and the Soviet Union. Ballistic missile defense is one of the main issues with regard to the weaponization of space. The initial research and development into ballistic missile defense systems used nuclear warheads in space in an attempt to intercept the threat ballistic missile. The interceptor missile of the system was usually designed to intercept the threat ballistic missile in the confines of space, although a layered approach was sometimes adopted which incorporated an endoatmospheric interceptor. This chapter examines the United States ballistic missile defense architectures and systems up to the signing of the Antiballistic Missile Treaty.

The Soviet Union's ballistic missile defense architecture is examined both up to the signing of the ABM Treaty and beyond, since the Soviet Union adhered to a different approach than the United States and continued to operate a ballistic missile defense system after the signing of the ABM Treaty. This chapter also addresses the ABM Treaty signed in 1972. It analyzes the treaty with regard to its limitations it placed on the United States and Soviet Union with respect to ballistic missile defense systems. The contentious issue of the interpretation of the treaty is examined with specific emphasis on the narrow and broad interpretations of the treaty which emerged with the United States' renewed interest in ballistic missile defenses in the form of the Strategic Defense Initiative.

The third chapter analyzes the Soviet Union's approach to military space development. It uses the theories of space power outlined previously to assess the Soviet Union's military use of space during the Cold War. This chapter examines the distinctive Soviet path to the weaponization of space with the Fractional Orbital Bombardment System (FOBS), an orbital weapon system capable of neutralizing the United States' early warning systems. The prospect of Russian cooperation during the period of the Global Protection Against Limited Strikes (GPALS) is assessed, along with Russian attitudes to joint early warning centers with the United States. The section progresses to examine the fate of the military space units in Russia since the breakup of the Soviet Union. This was an especially important issue since one of the Soviet Union's cosmodromes was located in Kazakhstan, a Soviet Republic and no longer part of Russia.

The latter part of the chapter examines the extent to which Russia has used cooperation with the United States to maintain its space industry in the face of reduced budgets. In particular Russia has developed a space launch industry which has been a rich source of hard currency.

The fourth chapter addresses the People's Republic of China's approach to military space. It traces the origins of China's military space programs and identifies the key personality Tsien Hsue-shen, involved in organizing and creating its military space program. The dual nature of China's civil space program combined with its military space activities is explored, along with the organizations that are involved in the process. These capabilities include photoreconnaissance, communications, meteorology, navigation, and electronic intelligence. China's space launch capabilities are outlined along with the PRC's geographical characteristics and its consequences for achieving specific orbits. China's distinctive track to space development including the military sphere and its international cooperation is highlighted. It is an extremely important characteristic of China's military space development that it relies heavily on its cooperation with other countries. The repercussions in the United States when this became apparent with U.S. industry created a highly politically charged environment which led to a Congressional investigation. This is examined with particular respect to the space launch market. China's interest in developing antisatellite weapons is the focus of their interest in the weaponization of space. The PRC has a number of programs to this end which are outlined, and appears to be seriously considering an active capability which could be used as a leveller against a potential adversary's space assets.

The fifth chapter analyzes the United States' and Soviet Union's antisatellite programs during both the Cold War and in the post-Cold War era. Initially the philosophy which believed antisatellite weapons had a destabilizing effect on the United States' relationship vis a vis the Soviet Union was espoused in the United States. The successive U.S. administrations' policies towards antisatellite weapons and technological systems that were under research and development are analyzed. The chapter additionally outlines the development of Soviet antisatellite weapons in terms of both its organizational structure and eventual testing and development of the capability. It analyzes the Soviet antisatellite testing methods to gain an insight into the strengths and operational capabilities of its programs. Toward the latter part of the Cold War in the late 1970s antisatellite arms control measures began to be seriously considered. These arms control discussions are analyzed. The end of the Cold War did not see the antisatellite issue disappear; indeed the issue has risen to the fore, especially in regard to the United States policy to seek control of space. The current U.S. antisatellite systems under development are outlined along with the policy rationale. Having analyzed the extensive Soviet antisatellite development during the Cold War the chapter addresses Russia's continuing work in this area.

The sixth chapter addresses the issue of weapons being based in space. The most likely weapon under consideration is the space-based laser for boost-phase ballistic missile interception. The chapter explores the scientific basis for space-based lasers including the chemical reaction which is required to produce the laser beam. The lethality of the space-based laser is considered with particular relation to the distance and intensity of the beam required to intercept the ballistic missile in flight. The orbital characteristics of the space-based laser are analyzed, and the basing of the satellites is addressed in relation to the orbit type and inclination required. The following section assesses the components that would be required to operate a space-based laser. The final section of this chapter outlines the technological programs that comprise the space-based laser. It examines the industries involvement with the United States Air Force and the Missile Defense Agency in the development of the space-based laser, and outlines the timetable for the eventual deployment of the system.

The seventh chapter examines the Revolution in Military Affairs and its relation to military space. The use of military space systems in the conflict in the Persian Gulf to expel Iraq from Kuwait was one of the first occasions that they contributed to the success of an air and ground campaign. The impact this had on the subsequent debate on the Revolution in Military Affairs (RMA) is analyzed. The impact this made had significant implications for those claiming a revolution in military affairs was underway. The RMA originated in the United States and the impetus and rationale for this is examined first.

The components of the RMA, precision strike, information warfare, and dominant maneuver are addressed in turn. The underpinning of these components by space systems is examined, and the consequences this may have are explored. The United States section concludes with an analysis of the factors driving the RMA and the concern in the United States of other countries embracing the RMA. In order to provide a comparative approach to the revolution in military affairs, attitudes of the People's Republic of China and Russia are also analyzed. In each case an attempt will be made to determine if a country has made a specifically unique approach to the RMA and adapted the concept to suit their individual national requirements.

The eighth chapter examines the United States' military space policy since the end of the Cold War. It analyzes President Clinton's two terms of office and the start of President Bush's administration with respect to missile defense policy and military space policy. The Clinton administration's period in office saw intense political maneuvering between Congress and the President over national missile defense plans. There were a number of congressionally initiated acts to instigate a program towards the building of a national missile defense system. This chapter outlines these architectures with regard to the use of an exoatmospheric interceptor, which is designed to intercept a threat ballistic missile in space. The impact of the Presidential Directives that were announced during the Clinton period on military space policy is addressed. The ensuing organizational changes with respect to military space that were implemented are explored.

The following section examines President Bush and missile defense policy. It analyzes the significance and rationale for the United States withdrawal from the Antiballistic Missile Treaty. The Commission to Assess the United States National Security Space Management and Organization reported during the first months of President Bush's administration. The impact this had on military space policy and the organizational changes it had on the space infrastructure are analyzed. In this section the space-based weapons that are under consideration are outlined, with particular attention given to space-based weapons against terrestrial targets. The chapter finally assesses the impact and contribution military space assets have made to recent conflicts. It examines the roles space assets made to the campaign in Yugoslavia, and the events in Afghanistan.

Notes

1. See James Canan, *War in Space* (New York: Harper and Row Publishers, 1982); Curtis Peebles, *Battle for Space* (Dorset: Blandford Press, 1983); Rip Bulkeley and Graham Spinardi, *Space Weapons: Deterrence or Delusion* (Cambridge: Policy Press, 1986) and Bengt Anderberg, Myron L. Wolbarsht, *Laser Weapons: The Dawn of a New Military Age* (New York: Plenum Press, 1992).

2. See William A. Owens, "The Emerging U.S. System of Systems," *Strategic Forum*, National Defense University, Institute for Strategic Studies, February 1996 and William A. Owens, "Revolution in Military Affairs? U.S. Vision for Future Warfare," *Royal Institute for International Affairs*, Conference Paper, May 21 and 22, 1997.

3. See Joan Johnson-Freese and Roger Handberg, *Space The Dormant Frontier: Changing the Paradigm for the 21st Century* (Westport, CT: Praeger Publishing, 1997) for a recent espousal of the sanctuary school of space power.

4. See Robert Giffen, "Space Power Survivability" in Uri Ra'naan and Robert L. Pfaltzgraff, ed., *International Security Dimensions of Space* (Hamden, CT: Archon Books, 1984) and C. S. Gray, "The Military Uses of Space: Space Is Not a Sanctuary,"

Survival, September/October 1993, C. S. Gray, "Space Power Survivability," *Airpower Journal*, Winter 1993.

5. See Karl Grossman, *Weapons in Space* (New York: Seven Stories Press, 2001), Steven Lambakis, *On the Edge of Earth: The Future of American Space Power* (Lexington: University Press of Kentucky, 2001).

6. See Edward Teller, *Better a Shield Than a Sword* (New York: Free Press, 1987). Teller was a strong proponent of the X-ray lasers and developed the concept of pop-up nuclear powered laser weapons, which were launched into space when required for ballistic missile defense.

Chapter One

The United States Approach to Military Space during the Cold War

This chapter examines the military uses of space during the Cold War period by the United States. The central thrust of the chapter is that both the Soviet Union and the United States formed a consensus on the military uses of space which sought to prevent it from being weaponized. However, in the 1980s this consensus began to weaken. In order to analyze this assertion this chapter will identify the theories of space power that emerged during the Cold War. The space policies announced by the United States will be examined, and then analyzed using the theories of space power.

During the latter part of the Cold War the consensus on the nonweaponization of space began to show signs of apparent weakening. Indeed as the following quote demonstrates, although there are no weapons in space this does not mean the issue was not considered during the Cold War.

> Since the space age began in earnest, dozens of space weapon system concepts have been seriously investigated and brought to varying degrees of development by both superpowers. From satellite interceptors to anti-ICBM networks to orbital bombardment systems, from conventional explosives to nuclear warheads to high energy beams and other exotic devices, these weapons vividly refute any notion of space as a sanctuary.[1]

The Strategic Defense Initiative and its subsequent reorientation to the Global Protection Against Limited Strikes (GPALS) was the epitome of the straining of this consensus. The final section of this chapter addresses the reorientation of the SDI program to GPALS and its effect on the consensus on the nonweaponization of space.

Military Space Power Theory

The following theories of space doctrine provide a useful analytical framework in which to view military applications of space during the Cold War. This section will outline the central tenets of the sanctuary school, the survivability school, the space control school, and the high ground school of space power. Having done this space policy will be explored with particular reference to the schools of thought from which it draws its theoretical underpinning.

The sanctuary view of space doctrine believes that the realm of space should not weaponized. The intrinsic value space provides for national security is that satellites can be used to examine within the boundaries of states, since there is no prohibited overflight for satellites as there is for aircraft. This enables arms limitation treaties to be verified by satellites in space serving as a national technical means of treaty verification.[2] Early warning satellites serve to strengthen strategic stability since they provide surveillance of missile launches which increases the survivability of retaliatory strategic forces. The sanctuary school argues that such is the importance of the functions of these space systems that space must be kept free from weapons, and antisatellite weapons must be prohibited, since they would threaten the space systems providing these capabilities.[3]

The survivability view of space doctrine believes that space forces are inherently less survivable than terrestrial forces. The origins of the survivability school can be traced to the late 1970s and the early 1980s with the testing and development of the Soviet antisatellite capability. This capability began to threaten U.S. space systems. The survivability school consequently argued that space forces must not be depended upon for providing various functions such as communications and surveillance in wartime because they may not survive.[4] However, such was the importance of space assets that they need to be protected as much as possible. This might imply the use of weapons in space. The United States during the 1980s pursued the goal of near-term military efficiency and committed itself to become dependent on space assets, in the knowledge that credible and effective threats would emerge.[5]

This leads on to the space control school which considers space as any other military theater and the military objective should be to seek control over the space environment. This is seen as analogous to concepts of air superiority and sea control. Although the space control school states that both defensive and offensive operations are likely to be conducted in space, it provides less focus on what specific purposes are served through space control.[6]

The high ground school of thought believes that space has the ability to be the critical factor in determining the outcome of a battle. The high ground school uses the analogy that the domination of the high ground ensures the domination of the lower areas.[7] It then follows from this that in the future, space forces will dominate terrestrial forces. This school of thought had its origins in President Reagan's Strategic Defense Initiative speech in that it advocates space-based ballistic missile defense. However, conceptually the high ground school envisions force application missions from space more than just for this purpose.[8]

The Development of U.S. Space Policy during the Cold War

There were several factors which accounted for the lack of U.S. interest in space during the aftermath of World War II. Firstly, the unknown potential of military space was unable to compete against the core missions of the military in the austere budget environment that followed the war. Secondly, many of the top scientific and military leaders believed that space-related technologies capable of making a contribution to national security, such as the ICBM, would not mature for many years. Thirdly, prior to the recognition that the Soviet Union was putting substantial resources into developing ballistic missile programs the U.S. was reluctant to give attention or funding to programs with unclear military potential.[9]

The primary goal of Eisenhower's space policy was to examine and exploit the potential of space to open up the closed Soviet state by using satellite reconnaissance. The second major goal was to design policies to create a new international legal regime which would legitimize satellite overflight for "peaceful purposes" including reconnaissance. The third major goal was to investigate space for scientific purposes. One of the most important aspects at this time was that the U.S. had to develop boosters capable of launching satellites or warheads over an intercontinental range which underpinned all of these goals.[10]

In the mid-1950s the main U.S. space policy goal was the development of reconnaissance. In support of this goal was the need to legitimize the operation of spy satellites. The satellite program, known as the WS-117L project included three programs on each of the three types of reconnaissance which would be used in the following decades: reconnaissance via recoverable film systems (CORONA); infrared surveillance for missile launch detection (MIDAS); and reconniassance via electro-optical systems (SAMOS).[11]

In 1957 none of the services had a comprehensive doctrine related to the potential military uses of space, with the development of space reconnaissance deemed the only acceptable aspect of space utilization. The creation of the National Aeronautics and Space Administration on 1 October 1958 saw added impetus for the civil route of the U.S. entry into space. The Eisenhower administration's policy of establishing space as an environment for peaceful purposes determined the de-emphasis of any other potential military missions in space.[12]

The reaction to the news on the 4 October 1957 that the Soviet Union had launched Sputnik I and had become the world's first spacefaring nation was to fundamentally shape U.S. space policy for several years. The administration renewed its calls for bringing other future developments in outer space under international control at the United Nations.[13] For the services Sputnik I meant that space was no longer a strategic backwater but could now offer a pathway to increased power and prestige. The Sputnik shock provided a rationale for the U.S. military to explore the requirement of an ASAT capability. Each of the services had proposed some form of ASAT proposal by November 1957.[14]

The launch of Sputnik I and the lack of international objection to its overflight effectively wrote into international law the right of satellite overflight of national territories.[15] The legality of satellite over flight was a policy goal of the Eisenhower administration, almost as much a priority as being first into space. The de facto achievement of this over flight right was significant for U.S. space policy towards its aim for satellite reconnaissance of the Soviet Union.

The Air Force moved during the initial period of the Sputnik shock to claim responsibility of U.S. military operations in space. This claim consisted of two interrelated parts: the development of the aerospace concept, and a high ground approach which asserted that space could make a critical contribution to national security. This was later reinforced with the first Air Force space doctrine announced by General Thomas D. White on 29 November 1957 which included the ideas that spacepower would prove as dominant in combat as the Air Force believed that airpower already was; there is one operational medium of aerospace since there is no distinction between air and space; and the Air Force should have operational control over all forces within this medium.[16] These assumptions held by the Air Force were in direct conflict with Eisenhower's space policy which followed the belief that space was a sanctuary for reconnaissance purposes.

President Eisenhower in 1958 established a special panel which wrote what later came to be known as the Purcell Report. This report reinforced Eisenhower's views on the militarization of space. Although the report had a wider remit on the scientific benefits of the exploration of space, it did endorse the military uses of space which were considered to be of specific utility. These included reconnaissance, communication, and weather forecasting.[17] The report's support for these passive military benefits of space also included a rejection of the notion of space weapons. This report was to establish the basic guidelines for the U.S. military exploitation of space. One of the first authors to discuss U.S. military satellite programs was Philip Klass.[18] This work gave a history of the development of various generations of reconnaissance satellites and was done with what little information was available at the time.

The military and the Air Force in particular, were encouraged by the determination of the Kennedy administration to close the supposed "missile gap" and renewed their efforts to increase the U.S. military presence in space against the acute tensions of the 1961-2 period. These hopes were largely dashed. By the end of the Kennedy administration the decision to cancel the Air Force's X-20 manned space vehicle and the concentration on the Apollo program meant that the U.S. was moving into space along a civil path.

It had, however, reassessed the sanctuary school of thought when Soviet statements and actions indicated that they were developing orbiting nuclear weapons. In May 1962, in order to counter the problem, Secretary of Defense McNamara tasked the Army with the modification of the Nike Zeus ABM for a future ASAT role.[19] The modified system, Program 505, was based at Kwajalein Atoll in the Marshall Islands. Each missile carried a nuclear warhead capable of destroying satellite targets. As the Soviets continued to pursue efforts toward an orbital bomb, pressure increased for the United States to develop an ASAT capability. In 1963 Kennedy approved Program 437, a ground-launched ASAT based on the Thor IRBM. Program 437, a forerunner to Program 505, was a rival Air Force program that received the ASAT mission when Program 505 was phased out. This issue will be explored in greater detail in a later chapter.

One of the major initiatives in this period was the negotiation by the Kennedy administration of the United Nations General Assembly Resolution 1884 (XVIII) on the 17 October 1963. This resolution called for the prevention of placing nuclear weapons or weapons of mass destruction in outer space.[20] This resolution laid the foundation for the Johnson administration to negotiate the Outer Space Treaty (OST) of 1967, which strongly influenced the development of subsequent military space doctrine. Some of the concerns regarding the OST were the possibilities for verification. The prohibition of military installations on the moon and other celestial bodies coupled with the banning of weapons of mass destruction from space, placed enormous restraints on the belief that space could openly serve as the high ground for deterrence or actual warfare at the strategic level.[21] The OST was one of the clearest signals that the U. S. civilian leadership did not believe that space held a great deal of military utility except as a sanctuary for reconnaissance satellites.

President Johnson continued the ASAT programs undertaken by the Kennedy administration, sharing the view that an ASAT was a hedge against Soviet orbital weapons. However, a report into ASAT weapons considered the use of ASATs against targets whether or not the orbital delivery weapons were introduced and advocated the U.S. ASAT capability as being able to enforce the principle of noninterference in space.[22] However, the Johnson administration did not share the view of the report's additional missions for the ASAT capability and instead reiterated that targeting Soviet satellites invited retaliation, and that the United States was more dependent on satellites. The Johnson administration did not seek to enhance the United States capabilities further than Program 437.

Shortly after entering office, President Nixon established a Space Task Group and tasked it to conduct a comprehensive review of the future plans of the U.S. space program. The tone of the report of the group which was published in September 1969, reflected the cost consciousness of the administration. It was announced that the Department of Defense would only be permitted to embark on new space programs when they could show it to be more cost effective to carry out the task in space.[23] The report's recommendations appeared to confirm what was already being undertaken in practice, such as the cancellation of the underfunded Manned Orbital Laboratory in June 1969.

The SALT I agreements comprised of the Treaty on the Limitation of Antiballistic Missile Systems and the Interim Agreement on the Limitation of Strategic Offensive Arms in May 1972 had considerable implications for military space policy. The primary impact on space policy as a result of these negotiations was on the central role for reconnaissance satellites to serve as a means of verification, and unclear restrictions for ABM systems. These agreements signaled that the U.S. military had made a departure away from the space control doctrine toward the sanctuary school.

The Ford administration in 1975 convened the Slicther Panel to review the military applications of space. The panel observed that the United States dependence on satellites was growing and that these assets were largely defenseless and prone to countermeasures.[24] This led to another panel to be initiated to analyze the vulnerabilities and to consider the need for an ASAT program. The panel known as the Buchsbaum panel considered that an ASAT capability would not enhance the survivability of U.S. satellites and that deterrence of attacks on satellites would be ineffective given the heavy dependence on space. This issue will be dealt with in a later chapter which focuses specifically on antisatellites. However, the blinding of U.S. satellites in 1975 and the resumption of Soviet ASAT testing led President Ford in 1977 to release National Security Memorandum 345 ordering the department of defense to develop an operational ASAT.[25]

The general lack of emphasis on military space issues can be seen from the fact that between the Kennedy and Carter administrations there were no major military space policy reviews undertaken at the NSC level.[26] There were however, two major policy statements on space during the Carter administration. These statements reflected the improvements in military technology and the increasing importance of space to the military. President Carter announced during a press briefing in March 1977 that he had proposed to the Soviet Union arms control measures to provide restrictions on an ASAT capability.[27] Carter's space policy can be seen as being dual tracked on the one hand he sought to establish a verifiable ban on ASAT systems, and on the other he pursued the development of an air-launched ASAT capability, with the Miniature Homing Vehicle contract awarded to the Vought corporation.[28] However, the signing of the SALT II Treaty on 18 June 1979 and subsequent invasion of Afghanistan by the Soviet Union, pushed the issue of an ASAT ban into the background.

On 11 May, PD-37, National Space Policy was signed by President Carter. This set out the twin track approach to ASAT developments along with the initiation of a long-term program to provide greater survivability for military space systems. It also stressed that the Secretary of Defense develop a plan to use civil and commercial space systems during declared national emergencies.[29] The Carter administration's top priority for its space policy remained the exploitation of space reconnaissance, but the increasing vulnerability of these systems and the need for the protection of these assets gradually led to a weakening of the sanctuary school of thought.

In 1981 President Reagan came to office. There had been little indication of the nature of the administration's military space policy during the election period and during the transition to office. The first space policy review was completed by the summer of 1982 and the National Security Decision Directive 42 set out the primary aims of U.S. space policy. These were not dissimilar to PD-37 in terms of diminishing satellite vulnerability, but a subtle shift in emphasis on ASAT policy occurred. Whereas the Carter administration had maintained that an ASAT arms control agreement was desirable, the Reagan policy was merely to "continue to study space arms control options."[30] There was also a shift in emphasis for the development of an ASAT capability to provide a means of deterring threats to U.S. space systems and denying any enhancement in the capabilities of the space-based forces of the potential enemies. A corollary to this was the requirement to develop a program capable of detecting threats to U.S. space forces and to provide a contingency in the event of such an occurrence.

The announcement of the Strategic Defense Initiative in March 1983 set out a research and development program into the feasibility of utilizing space for strategic defense. This, coupled with the Challenger disaster of January 1986 led to a revised policy on U.S. space policy in January 1988. This set out four basic requirements for U.S. space policy, were specified as follows:

1) deterring, or if necessary, defending against enemy attack;
2) assuring that forces of hostile nations cannot prevent our own use of space;
3) negating, if necessary, hostile space systems; and
4) enhancing operations of United States and Allied forces.[31]

This directive builds upon the foundations of military space doctrine which includes: space support, force enhancement, space control and force application. The space support mission mandated the Department of Defense (DOD) to maintain launch capability on both coasts and to enhance the robustness of its satellite control capability. For force enhancement, the DOD was to develop space systems and plans to support operational forces at all levels of conflict. In the space control area, DOD was directed to develop an integrated combination of antisatellite survivability, and surveillance capabilities. And finally, under force application, the DOD was to conduct research, development, and planning to be prepared to deploy space weapons systems for strategic defense should national security conditions dictate it.

The Evolution Of The Strategic Defense Initiative Program

The original task of the Strategic Defense Initiative (SDI) was to research the feasibility of a missile defense capable of breaking up a determined Soviet nuclear attack on the United States that could consist of thousands of nuclear weapons. This Phase I of the SDI program was to ensure that a large percentage of these nuclear weapons would be destroyed. Although the military requirements established for SDI in the Reagan Administration are classified, it has been reported that Phase I would have been able to shoot down 30 percent of all warheads fired in a first strike[32] and 50 percent of the warheads carried on the SS-18 Satan missile, whose combination of accuracy and yield made it the most dangerous counterforce threat in the Soviet arsenal.[33] This would mean that a Soviet war planner could not successfully plan for a first strike, since he would be unsure of how effective his first strike could be. Phase I was to be only the start of a larger defense system, to meet possible changes in the Soviet threat.[34]

The Department of Defense began a study into the technological feasibility of missile defenses. This study formed a panel commonly known as the Fletcher Panel. The major recommendation of the Fletcher study was for a long-term research and development program on ballistic missile defense.[35] No specific BMD systems were selected for ultimate deployment, but promising new technologies and systems were identified for research. Decisions about further research and development for deployment would be made after an initial five years of study. The technologies study team placed its emphasis on a long-term program to research and develop a multitiered defense that would provide significant damage limitation. It believed that a credible defense would have to have a low leakage of warheads, but no actual number was defined.[36] The study de-emphasized short-term, narrowly defined program elements.

The Fletcher Panel estimated that the research and development programs could last ten to twenty years to enable critical technological problems to be solved and begin deployments.[37] In essence the Fletcher Panel approach overly stressed performance standards to the detriment of the deployment of limited but potentially effective defenses. A major flaw of the Fletcher Panel's report was that it called for research into BMD technology to continue until it was possible to make and deploy in its entirety a nearly flawless defensive system against very large scale and very sophisticated attacks. One critic commented

> President Reagan's Fletcher Panel crafted gold-plated definitions of what is required for ballistic missile defense. Moreover, the Fletcher Panel stated plainly that even these definitions might be made more demanding to take account not of what intelligence learns the Soviets are doing but rather the American technician's own evolving notions. In bureaucratese, such changing of standards is known as the "responsive threat." Thus was the SDI established as a program of research without logical end.[38]

The Future Security Strategy Study, also known as the Hoffman Panel, established at the same time as the Fletcher Panel, was given the assignment to study the implications of strategic defenses vis-a-vis the relationship between the United States and the Soviet Union. The panel found that strategic defenses, even if not perfect, were not inconsistent with the goal of helping to stabilize the U.S.-Soviet military relationship.

With this opinion, the Hoffman Report differed from the Fletcher Report. The Hoffman Report advised early deployment of partial strategic defenses, even if they were not the highly capable multilayered defenses envisioned by President Reagan and the Fletcher Panel. Even limited defenses, it said, could greatly enhance deterrence by denying the Soviet Union at least some of its military objectives. In short, the Hoffman Report called for a healthy mix of offense and defense to enhance the United States' deterrent. The Fletcher Report, on the other hand, advised waiting until a highly capable system could be deployed all at once.[39]

The SDI Mission

The SDI program from its conception in 1983 eventually changed its focus significantly. At first, it focused on the threat of a massive Soviet attack, but by 1991 it had switched to protection against much more limited strikes from anywhere on the globe. During each of these mission periods, there were significant changes to the program, based on policy reviews and decisions by the President, the Secretary of Defense, and the Congress. The first phase and change to the SDI program began with the development of the technology period up until late 1987. The next major step that followed this period was Phase I with the Space-based Interceptor, and the final step before the switch to GPALS was Phase I with Brilliant Pebbles.[40] Despite all these changes these steps were all known as Phase I, since they were intended to defend against a massive Soviet attack, although their architectures were subject to changes.

Phase I, the defense against a massive Soviet attack phase began with its creation in 1983 and lasted through until 1990. In accordance with directives from the President, the Secretary of Defense chartered the Strategic Defense Initiative Organization (SDIO) in 1984 to research and develop a set of technologies supporting concepts for Ballistic Missile Defense (BMD).[41] SDIO was to support a decision to be made in the early 1990s on whether to begin developing BMD for deployment. Initial deployments were to contribute to strategic defense and move the United States toward a goal of eliminating the strategic nuclear missile threat. SDI was also to protect options for near-term deployment in case of a Soviet deployment in violation of the Antiballistic Missile (ABM) Treaty.[42]

The SDI program was to be treated as a research program until the early 1990s, when a decision would be made to decide whether to develop and deploy an initial capability. The SDIO was developing a wide range of key technologies for sensors, kinetic kill weapons, and directed energy weapons. As President Reagan stated, "the SDI program was to provide to a future president and a future Congress the technical knowledge required to support a decision in whether to develop and later deploy advanced defensive systems."[43] In 1987 Phase I became subject to the oversight of the Department of Defense's formal acquisition process.

In the fall of 1986 a Phase I national missile defense design was developed. The concept of phased deployment was to "develop and deploy militarily useful increments of capability" that would also add to arms control negotiating leverage for reductions in offensive weapons.[44] If the Soviets responded favorably to arms reduction proposals the phased deployment proposals could be modified. There were three phases. The first phase aimed at denying Soviet initial strike objectives, along with the ability to blunt follow-on strikes, which would complicate Soviet attack options and defeat limited attacks and accidental launches. The early follow-on phase included directed energy systems and active discrimination sensors. The final phase, the late follow-on phase, included advanced directed energy weapons and support technologies. The latter two phases would lead to highly effective, multilayered defenses.[45]

The general outlines of the Phase I and follow-on deployment concepts were approved by President Reagan in December 1986. Phase I emphasized the space-based elements as being of critical importance to countering the Soviet proliferation of offensive missiles. The White House also called SDI "a main inducement for the Soviets to negotiate for deep cuts in offensive arsenals."[46] President Reagan declined Soviet demands to confine SDI to laboratory research.

The Defense Acquisition Board's (DAB) review in September 1987 led to the recommendation of selected Phase I elements. The selected Phase I elements were:

- Boost Surveillance and Tracking System;
- Ground-Based Surveillance and Tracking System;
- Space-Based Surveillance and Tracking System;
- Space-Based Interceptor;
- Exoatmospheric Reentry Vehicle Interceptor System;
- Ground-Based Radar;
- Battle management/command, control, and communications;
- System engineering and integration and launch[47]

In September 1987, Secretary Weinberger approved the recommendation by the DAB that Phase I concepts and technologies, called the Phase I Strategic Defense System, enter the validation section of the acquisition process.[48] The advanced technologies for follow-on phases were to enter demonstration and validation prior to full-scale development of Phase I. The need to lower costs and resolve effectiveness issues such as survivability, vulnerability, and sensor performance meant that Phase I underwent continual design and renewal.[49] Success in this endeavor saw the cost estimates of a Phase I defense of the United States reduced from an original June 1987 DAB estimate of $145.7 billion, to $115.4 billion in June 1988, then $69.1 billion in September 1988, and to $55.3 billion by November 1989.[50] These reductions came about by successive redesign of the system elements, reductions of support costs, and changing cost-estimating models.[51]

After Phase I was proposed, the SDIO began investigating a new, innovative space-based interceptor, known as "Brilliant Pebbles." These were to be a constellation of up to thousands of individual interceptors, each with its own surveillance capability and enough power to operate autonomously, within its own field of vision. Brilliant Pebbles was a competitor to the Space-Based Interceptor design concept, which was to house several interceptors together in a large "garage" or carrier vehicle. Brilliant Pebbles responded to the DAB concerns over the high cost of the Space-based Interceptor "garage" or carrier vehicle. Primarily, however it allayed concerns relating to the survivability of the Space-Based Interceptor garage. Brilliant Pebbles was subjected to several technical feasibility reviews in 1989.

President Bush, upon entering office in 1989, directed a National Security Review which was headed by Ambassador Henry F. Cooper.[52] This review, which was completed in spring 1990, endorsed the concept of Brilliant Pebbles and recommended its innovative approach be applied to the rest of the SDI Phase I architecture. In testimony before Congress in April 1990, the Director of the SDIO announced that Brilliant Pebbles had replaced both the Space-Based Interceptor and the Boost Surveillance and Tracking System in Phase I. In June 1990, the Undersecretary of Defense for acquisition endorsed the changes.

In November 1990 the SDIO recommended revisions, which included the replacement of the Space Surveillance and Tracking System satellites, with smaller, highly distributed Brilliant Eyes satellites. These satellites were drastically smaller than the previous missile tracking satellites which would have made them easier to defend from attack. The Endo-Exoatmospheric Interceptor was introduced as a competitor to the exoatmospheric Ground-Based Interceptor and design changes were made to the Ground-Based Radar, redesignating it the Ground-Based Radar-Terminal.[53]

During the Reagan administration, it was the White House which set the most ambitious plans for military space rather than the Pentagon, a reversal of the formulation of military space policy under the Eisenhower and Kennedy administrations.

President Bush's Global Protection against Limited Strikes System

The change in the U.S.-Soviet relationship and more influentially the breakup of the Soviet empire caused a reevaluation of the purpose of the SDI program. This reevaluation was to be widespread throughout U.S. military strategy, as President Bush announced that U.S. military strategy was to be significantly altered from fighting a global war against the Soviet empire to fighting regional conflicts against a variety of potential aggressors.[54] It is in this context that the SDI program was reoriented.

The SDI program was to be "refocused on providing protection from limited ballistic missile strikes, whatever their source."[55] The smaller-scale SDI would aim to provide high protection against a smaller number of missiles. It would not assume that the Soviet Union was the aggressor, or that the United States was the target. The system would be capable of protecting not just the United States, but military forces overseas and allies.[56]

> GPALS pares back America's SDI plans to meet the fiscal and military requirements of the 1990s. Unlike Reagan's SDI program, designed to disrupt a massive Soviet surprise attack involving thousands of incoming missiles, GPALS will give America—and its allies—a near-perfect defense against limited or perhaps accidental attacks by up to 200 missile warheads. GPALS then cuts the proposed costs of SDI from $53 billion to $41 billion over ten years. This puts SDI well within the cost-range of other important defense programs—less than the Air Force's B-2 Stealth bomber and comparable to the mobile Midgetman missile system.[57]

GPALS was designed to provide near-perfect protection against smaller strikes, potentially from a Third World foe or a fragmented Soviet Union. It was to be able to defend missile strikes of up to two hundred warheads[58] aimed at the U.S. from anywhere in the world with near 100 percent confidence.[59] This stands in contrast to the Phase I SDI mission which was developed under the Reagan Administration.

In 1991 General Colin Powell, Chairman of the Joint Chiefs of Staff, in a statement before the Committee on Armed Services, said that the Pentagon officially still retained the military requirements for a full Phase I SDI system as a long term goal for the U.S. ballistic missile defenses.[60] If necessary, GPALS could have been expanded through the deployment of additional interceptors to meet Phase I requirements.[61]

During the period 1989-1990 the Department of Defense and SDIO reacted to new forces affecting SDI. These new forces were the innovations in the Brilliant Pebbles concept and the changes in Soviet and third-world threats. The events in the world during this period led to a reexamination of the policy and technical goals of the SDI program. This led to an Office of the Secretary of Defense study of "the strategy and technical feasibility of global protection against limited strikes" in the spring and summer of 1990.

In January 1991, President Bush refocused the SDI program to deal with accidental or unauthorized launches of ballistic missiles and with deliberate attacks of limited scope. As President George Bush declared in his State of the Union Address:

> Looking forward, I have directed that the SDI program be refocused on providing protection from limited ballistic missile strikes, whatever their source. Let us pursue an SDI program that can deal with any future threat to the United States, to our forces overseas and to our friends and allies.[62]

Whilst the threat for GPALS was less technically stressing, the mission of near-perfect protection put additional stresses on designs. As the report to the Chairman on Governmental Affairs in the Senate argued:

> High levels of protection require near perfect system performance in detecting, discriminating, and tracking targets; in battle management, command, control, and communications functions; and in intercepting and destroying targets.[63]

The Missile Defense Act in 1991 changed the shape and priorities of the GPALS program.[64] The act set goals for the early deployment of advanced theater missile defenses and the initial site for the defense of the United States against limited attack. Congress gave the Department of Defense 180 days to develop a plan to meet its mandate for early deployment. It also mandated that Brilliant Pebbles space-based interceptors would not be part of initial planned deployments, but be pursued in "robust" research and development.[65]

In November 1991, the SDIO briefed the DAB's coordinating committee that the Theater High Altitude Area Defense program, including the Theater Missile Defense Ground-Based Radar, had high cost and schedule risks. SDIO was requested by the coordinating committee to develop acquisition strategy options to reduce risks. The Army and the SDIO subsequently modified the program, which consequently led to the DAB approving a milestone I entry[66] into demonstration and validation. DOD reviewers identified concurrency risks in meeting the Congress" early fielding goals for an initial, single-site national missile defense system. After considering the DOD's assessment, Congress amended the Missile Defense Act in 1992, delaying the proposed fielding date. The

1992 Act continued the restrictions on the deployment of space-based interceptors.

One of the key elements of the refocused SDI program was the increased priority for the theater missile defense programs. The experience in the Persian Gulf with the Patriot missile focused more attention on this priority. One of the objectives of the program was to focus on near-term deployment of improved theater missile defense systems. This is an area where cooperation with allies could be expanded.

The Case for Space-based Assets

The issue of why space has to be utilized was a key issue in regard to the refocusing of the SDI program. The issue was not space versus ground; even with an emphasis on ground systems space elements were required. At a minimum, space-based sensors made ground systems more effective. The issue was not whether to utilize space; space-based sensors would be required. The real issue was whether or not to use space-based weapons. There would be many advantages to using such space-based systems, particularly as the threat matured and improved over time.

Space-based weapons would always be in position. They could defend and offer protection against threats to forces arriving in-theater before theater commanders have had the opportunity to establish their own theater defense capabilities. Space offers broad area coverage to protect "a wide array of assets from a system based in space rather than having to protect each of those assets with their own individual ground-based systems."[67] Space-based weapons were a key hedge against a possible resurgent threat.

The need for highly effective defenses placed a premium on the ability to take multiple shots against ballistic missiles, including shots from space. A layered defense, combining surface- and space-based interceptors (SBI), provided the highest confidence in achieving protection for the United States against limited missile threats. Space-based interceptors would constitute the initial layer of a multilayered defense. They offer a defensive tier, with warning, command, and control, and intercept technologies that are independent of those dedicated to the surface-based layer.[68]

It was argued that space-based defense could provide multiple early engagements, well away from the defended targets. One of the lessons of the Gulf War was the importance of intercepting at distances and altitudes sufficient to prevent portions of a ballistic missile or its warheads from striking the intended target. Space-based interceptors could have mitigated one of the limitations associated with the Patriot missiles during Operation Desert Storm. Intercepts could have taken place above the atmosphere and debris from destroyed missiles could have been less harmful by the time of impact.[69] The destruction of nuclear, chemical, and biological weapons above the atmosphere would be important to prevent fallout over military or civilian target areas and the dispersal of chemical and biological weapons.

Space-based interceptors would provide global defensive coverage and could contribute to U.S. military strategy for regional conflicts. This would be particularly valuable for effective defense of U.S. forward-based and expeditionary forces because the location and timing of regional conflicts cannot be predicted, and may occur with little warning. Space-based interceptors would assist in protecting U.S. forces that must be deployed rapidly abroad. The forward-deployed forces would increasingly be operating within range of ballistic missile threats. As DeBiaso argues:

> In such a contingency, where an adversary might attempt to oppose the initial build-up of U.S. and allied forces, with ballistic missile strikes against ports, airfields, and early arriving troops, space-based interceptors could offer protection before surface-based interceptors were in place, thereby helping to maintain stability during a period of escalation and mobilization.[70]

It was argued that basing defenses in space would also provide a cost-effective protection for U.S. forward-based and expeditionary forces, and reduce the overall requirement for surface-based interceptors and their associated level of manpower and logistic support. During Operation Desert Shield, more than 450 C-141 equivalent air sorties were flown to transport ground-based missile defenses into the theater.[71] Space-basing in combination with new generation theater missile defenses would reduce such logistical burdens and also ease the overseas basing issues associated with deploying large numbers of surface-based interceptors globally.

The global coverage offered by space-based interceptors could provide a unique capability to defend multiple theaters simultaneously. This would be particularly important to the conduct of U.S. military operations as the threat of ballistic missiles extends beyond any single theater, especially in areas where ground-based defenses might not be deployed. For example, in the event of a crisis in the Middle East, space-based interceptors could provide protection to vulnerable U.S. and allied targets in adjacent theaters, such as cities, staging points or forces, necessary for operations in the primary theater.[72]

In the same way U.S. forces deployed overseas would benefit from the combination of space- and surface-based defensive systems, so would U.S. allies. The deployment of space-based interceptors would provide an initial defense tier complementing the allies" own ground-based defenses, resulting in protection against the entire range of threats. As DeBiaso argues:

> U.S. space-based interceptors could ease the burden of allied costs for theater missile defenses, thereby increasing the incentives for allies" investment in their own ground-based defenses. These allied theater missile defense systems could, in turn, provide additional coverage for U.S. forward deployed and expeditionary forces, especially against short-range missile attack. More broadly, deploying defenses in space should help demonstrate U.S. support for its allies by providing a unique military capability, despite reductions in forward deployed nuclear and conventional forces.[73]

The GPALS 1992 Architectural Design

The refocusing of the original SDI mission of destroying around half of a mass raid involving several thousand reentry vehicles launched out of the Soviet Union to protection of limited strikes meant a new architectural design. A component of GPALS would be transportable defenses that could be moved into a theater or region, if and when a hot spot might develop. In places that have continuing hotspots, defenses could be deployed indigenously, such as the Arrow system developed jointly with the Israelis.[74]

The ground-based element of GPALS was a defense against strategic ballistic missiles to be deployed in the United States. This ground-based system includes a satellite sensor, Brilliant Eyes. This sensor would improve the effectiveness of theater defenses as well as defenses against longer-range strategic ballistic missiles. The inclusion of Brilliant Eyes reiterates an important aspect about space versus ground-based defenses. Ground-based defenses require space-based sensors if they are to reach their potential.

The Ground-Based Radar (GBR) was much smaller and more mobile than in the previous SDI architecture. Its development built on the program of the smaller GBR employed in the theater missile defense system. Other elements of the ground-based system included two interceptors. The Exo-atmospheric Interceptor, (E2I) was to perform its intercepts high in the earth's atmosphere, after the reentry vehicles were able to be distinguished from lighter decoys since the atmosphere causes distinct deceleration characteristics.[75]

The final component of the GPALS system was the space-based interceptor called Brilliant Pebbles. Each Pebble was to be an autonomous interceptor which could act independently once it had been authorized. "It basically looks and sees the ballistic missiles when they rise from their silos, or, in the case of a Scud, from a mobile launcher. At the appropriate time, it drops its "life-jacket" and proceeds to maneuver into the oncoming path of the threat ballistic missile—or during the midcourse phase, of a reentry vehicle transiting space."[76]

It was a misconception that Brilliant Pebbles could not be employed effectively to counter theater ballistic missiles. If the range was greater than a few hundred miles, normal minimum energy trajectories would carry the RVs above the earth's atmosphere and there would be time to intercept them from space, using Brilliant Pebbles.[77] Such a system could be employed to counter the Scuds launched out of Iraq into Tel Aviv and Riyadh. The debris from such intercepts would probably burn up when reentering the earth's atmosphere rather than fall on city streets.

GPALS would have consisted of antimissile systems developed by the SDI program, but GPALS would have required fewer of them. The number of space-based interceptor missiles was reduced from over 4,000 in the Phase I plan to 1,000 in the GPALS plan. The number of ground-based interceptors was halved from 1,600 to 800.[78] If deployed in the proper orbit, Brilliant Pebbles could intercept ballistic missiles with ranges from about 300 miles to intercontinental distances. However missiles with ranges below 300 miles do not climb above 62 miles and thus do not reach altitudes high enough to become vulnerable to space-based Brilliant Pebbles interceptors.[79] During the Persian Gulf War, Iraq's al-Hussein and al-Abbas missiles, with ranges of 375 and 550 miles, would have been vulnerable to Brilliant Pebbles.[80]

Space-based weapons remained essential since they would have been far more effective than ground-based interceptors against missiles with multiple warheads. Ground-based interceptors must discriminate between warheads and decoys and then attack each of the warheads individually in space or as they reenter the earth's atmosphere closing in on their targets. Brilliant Pebbles would not have needed to do this because it would destroy the one missile carrying the warheads and decoys.

Any warheads that slip through the Brilliant Pebbles net in space would have been intercepted by ground-based interceptors. Two ground-based interceptors were under consideration. The Ground-Based Interceptor (GBI) and the Exoatmospheric/Endoatmospheric Interceptor or E2I. The GBI was based on technology developed through the Exoatmospheric Reentry Vehicle Interceptor Subsystem (ERIS) test program.

A test version of ERIS intercepted and destroyed a U.S. Minuteman I dummy warhead in space in January 28, 1991.[81] The targeted Minuteman I was launched from Vandenberg Air Force Base, while the test version of ERIS was launched from Kwajalein Atoll in the Pacific Ocean. Ground-based interceptors like ERIS, would attack enemy missile warheads in space before they reenter the atmosphere.

The E2I was based on technology developed through the High Endoatmospheric Defense Interceptor (HEDI) program. E2I was designed to intercept and destroy enemy warheads after they reentered the atmosphere. It generally would attack only after the earth's atmosphere had stripped away the decoys. The challenge the E2I faced was ensuring that its onboard sensor would find the target warhead and direct the interceptor against it. This is tougher to accomplish inside the atmosphere than above it, since the speed of the incoming missile creates friction with the atmosphere that then creates extremely high heat. This heat distorts the view seen by the E2I's sensor as it "looks" through its window.

GPALS depended on ground-based as well as space-based sensors to track ballistic missiles in flight. The ground-based sensors would relay essential targeting information to the interceptor missiles so that they could locate and destroy enemy warheads. Two ground-based sensor systems were included in the GPALS system.

The first of these systems was the Ground-Based Radar (GBR), which tracks missile warheads in the latter stage of their flight in space and inside the atmosphere as they close on their targets. The Ground-Based Radar was particularly useful in tracking missiles that have shorter times of flight, such as submarine-launched ballistic missiles (SLBMs), since it had the ability to process radar information quickly and provide it to commanders.[82] The Ground-Based Radar system was designed to be mobile and was envisioned to be deployed on railcars to make it less vulnerable to enemy strikes.

The second of the GPALS ground-based sensor system was the Ground-based Surveillance and Tracking Systems (GSTS). This was a heat-sensitive sensor mounted on a rocket. Upon early warning of a missile strike, the sensor would be launched into space to scan for incoming warheads beyond the range of the ground-based radar.[83] The GSTS system was to play an important role in distinguishing between real warheads and decoys.[84]

The Theories of Space Power Underpinning U.S. Space Policy

Space policy during the Eisenhower administration followed the sanctuary view of space. The focus on utilizing satellites for reconnaissance purposes combined with the reluctance to countenance the protection of these satellites by means other than international law, and a treaty-based approach to establish legal overflight of these satellites fits into the sanctuary view as outlined previously. Although the Air Force aerospace doctrine advocated by General White departed from the sanctuary view, this met strong resistance from the Eisenhower administration.

During the Kennedy and the subsequent Johnson administration the sanctuary school view of space was visibly highlighted with the culmination of the Outer Space Treaty which prohibited weapons of mass destruction being placed in orbit. This significantly curtailed the high ground view of space which sees space as a place from which Earth could be dominated, presumably with the placing of nuclear weapons in space. Also, the policy of utilizing satellites for reconnaissance purposes continued to follow the sanctuary school of space.

The Nixon administration's most significant space policy act was the signing of the Antiballistic Missile Treaty in 1972. This placed limits on ballistic missile defense and hence had implications for the high ground of military space theory which sees ballistic missile defense in space as an integral part of the military utility of the "high ground." Also, the SALT I Treaty for the first time advocated a NTM (national technical means) code for reconnaissance satellites as a means of monitoring arms control agreements. This action followed a sanctuary view of space, as a means of using space for peaceful purposes, that is, as a way of strengthening strategic stability, since it was believed that no side would cheat if there was a reasonable chance that the other side had a means of verifying whether they were adhering strictly to the terms of the treaty.

The Carter administration pursued a policy on similar lines which followed the sanctuary school of space theory. However the feeling of satellite vulnerability gave some credence to the vulnerability school of space power. President Carter, faced with possible satellite vulnerability, ordered a policy of research and development into a possible antisatellite capability. The research and development of such a capability would however lend itself towards a space control theory of space power. However, it can be assumed that the administration was developing an ASAT capability as a prelude to an ASAT ban, using an ASAT capability as a negotiating tool with which to bargain. However, an ASAT ban was to prove a difficult treaty to negotiate and was never realized.

The Reagan administration's space policy was a dramatic departure from the sanctuary school of space power. The development of an ASAT while maintaining that an ASAT treaty was undesirable lends itself toward the space control view of space power, but combined with the announcement of SDI, gravitates Reagan's space policy toward the high ground. Indeed, the development of a space-based ballistic missile defense was one of the fundamental tenets of the high ground view of space power. The announcement of a military space doctrine which valued space support, force enhancement, space control, and force application leaned heavily towards the space control view of space power. However, to summarize the Reagan administration's space policy can be classified as following the space control view of space power, but with a view towards the future of a high ground view of space, with the research and development of space-based ballistic missile defense capabilities.

The Eisenhower administration's space policy was highly secret and had a heavy focus on space being a sanctuary for spy satellites. The sanctuary school of thought continued throughout the Kennedy administration right up until the Ford administration. The Ford and Carter administrations saw the revival of interest in ASAT issues, but still the prevailing school of thought was the sanctuary school. The Reagan period saw a significant shift away from the sanctuary school of thought with the SDI speech, which can be characterized as the "high ground" school of space theory. Despite the rhetoric of the SDI speech, a more sober analysis of the Reagan administration would better be classified as a move toward the space control school of thought. This can be evidenced by the four basic tenets of U.S. space policy as set out in 1988 which covers space support, space enhancement, space control and force application. However, throughout most of the Cold War period the United States viewed space as a sanctuary free from the deployment of weapons.

The Bush administration, facing an altered geostrategic environment, especially vis-a-vis the United States-Soviet relationship, redirected the SDI mission to the GPALS mission. This redirection of the SDI mission clearly weakened the notion of space being free from weaponization, as the space architecture envisioned space components for the interception of ballistic missiles. The rationale for the use of space components for ballistic missile defense was that they provided a layered approach which could allow multiple early engagements away from the defended areas. However, not unlike the original SDI program the use of space for ballistic missile interception was left primarily a research and development program.

Notes

1. Nicholas L. Johnson, *Soviet Military Strategy in Space* (London: Jane's Publishing Company, 1987), 9.
2. David E. Lupton, *On Space Warfare: A Space Power Doctrine* (Alabama, Maxwell Air Force Base: Air University Press, 1988), 35.
3. Peter Hays, *Struggling Towards Space Doctrine: U.S. Military Space Plans, Program, and Perspectives During the Cold War*, Ph.D. Thesis, Fletcher School of Law and Diplomacy, 1994, 22.
4. Lupton, 36.
5. C. S. Gray, "The Military Uses of Space: Space Is Not a Sanctuary," *Survival*, Volume XXV, Number 5, September/October 1983, 197.
6. Hays, 24.
7. Lupton, 36.
8. Hays, 25.
9. Hays, 62.
10. Hays, 63.
11. William E. Burrows, *Deep Black: Space Espionage and National Security* (New York: Berkley Books, 1986), 80.
12. Hays, 97.
13. Walter A. McDougal, *The Heavens and the Earth: A Political History of the Space Age* (New York: Basic Books, 1985), 127.
14. Paul B. Stares, *The Militarization of Space* (Ithaca: Cornell University Press, 1985), 49.
15. Joan Johnson-Freese and Roger Handberg, *Space the Dormant Frontier: Changing the Paradigm for the 21st Century* (Westport, CT: Praeger Publishing, 1997), 46-47.
16. White's speech is reprinted in Eugene M. Emme, ed., *The Impact of Air Power: National Security and World Politics* (Princeton, NJ: D. Van Nostrand, 1959), 496-501.
17. Stares, 46.
18. Philip Klass, *Secret Sentries in Space* (New York: Random House, 1971)
19. David W. Ziegler, *Safe Heavens: Military Strategy and Space Sanctuary Thought* (Maxwell Air Force Base, Alabama: Air University Press, June 1998), 11.
20. Stares, 90.
21. Hays, 224.
22. Ziegler, 12.
23. Stares, 159.
24. Ziegler, 13.
25. Ziegler, 14.
26. Hays, 264.
27. Stares, 181.
28. Stares, 184.
29. *President's Space Report, 1978*, 99-100, quoted in Hays, 266.
30. "White House Fact Sheet Outlining Space Policy," quoted in Stares, 218.
31. Department of Defense, *Department of Defense Space Policy* (Washington, D.C.: March 10, 1987, 2-5 quoted in Dana Johnson, *The Evolution in Military Space Doctrine: Precedents, Prospects, and Challenges*, Ph.D. Thesis, University of Southern California, December 1987, 288.

32. Hildreth, Stephen A., *The Strategic Defense Initiative: Issues for Phase I Deployment*, CRS Issue Brief (Washington, D.C.: Congressional Research Service, 1990), 4.
33. Hildreth, 40.
34. J. D. Crouch II, 'SDI and Securing Western Freedom," *Laissez-Faire*, vol. 1 no. 4, Summer 1992, 17.
35. Franklin A. Long, Donald Hafner, and Jeffrey Boutwell, eds., *Weapons in Space* (New York: Norton, 1986), 50.
36. Long et al, 70.
37. Long et al, 94.
38. Angelo Codevilla, *While Others Build* (New York: The Free Press), 1988, 9.
39. Keith B. Payne, *Strategic Defense: "Star Wars" in Perspective* (Lanham, MD: The Hamilton Press, 1986), 19-21.
40. Report to the Chairman, Committee on Governmental Affairs, U.S. Senate, *Ballistic Missile Defense Evolution and Current Issues* (U.S. General Accounting Office, July 16, 1993), 22.
41. GAO Report, 25.
42. GAO Report, 25.
43. President Reagan quoted in Report to the Chairman, Committee on Governmental Affairs, U.S. Senate, *Ballistic Missile Defense Evolution and Current Issues* (U.S. General Accounting Office, July 16, 1993), 25.
44. GAO Report, 25.
45. GAO Report, 26.
46. Weinberger, Casper W., *Fighting for Peace: Seven Critical Years in the Pentagon* (New York: Warner Books), 1990, 324.
47. Report to the Chairman, Committee on Governmental Affairs, U.S. Senate, *Ballistic Missile Defense Evolution and Current Issues* (U.S. General Accounting Office, July 16, 1993), 27.
48. GAO Report, 27.
49. Particularly the challenge of discriminating targets in the midst of countermeasures designed to confuse SDI sensors.
50. Report to the Chairman, Committee on Governmental Affairs, U.S. Senate, *Ballistic Missile Defense Evolution and Current Issues* (U.S. General Accounting Office, July 16, 1993), 28.
51. GAO Report, 28.
52. Ambassador Cooper subsequently served as Director of the Strategic Defense Initiative Organization, from July 1990 to January 1993.
53. Report to the Chairman, Committee on Governmental Affairs, U.S. Senate, *Ballistic Missile Defense Evolution and Current Issues* (U.S. General Accounting Office, July 16, 1993), 29.
54. Remarks made by the President at the Aspen Institute Symposium in Aspen, Colorado, August 2, 1990.
55. Stephen J. Hadley, and Henry Cooper, *Briefing on the Refocused Strategic Defense Inititative*, February 12, 1991.
56. J. D. Crouch II, "SDI and Securing Western Freedom," *Laissez-Faire*, Vol. 1 No. 4, Summer 1992, 18.
57. Baker Spring, "For Strategic Defense: A New Strategy for the New Global Situation," *Heritage Foundation*, April 18, 1991, 2.
58. The figure of 200 warheads was used since this represents the number of warheads on a Russian SSBN.

59. Spring, 4.

60. General Colin Powell, *Statement of the Chairman of the Joint Chiefs of Staff Before the Committee on Armed Services*, U.S. House of Representatives, February 7, 1991, 9.

61. Heritage Foundation briefing by Administration officials on GPALS on February 11 and February 21, 1991, quoted in Baker Spring, "For Strategic Defense: A New Strategy for the New Global Situation," *Heritage Foundation*, April 18, 1991, 5.

62. George Bush, *Presidential State of the Union Address*, January 29, 1991.

63. Report to the Chairman, Committee on Governmental Affairs, U.S. Senate, *Ballistic Missile Defense Evolution and Current Issues* (U.S. General Accounting Office, July 16, 1993), 30.

64. The Missile Defense Act of 1991 will be discussed further in the next chapter.

65. Report to the Chairman, Committee on Governmental Affairs, U.S. Senate, *Ballistic Missile Defense Evolution and Current Issues*, (U.S. General Accounting Office, July 16, 1993), 31.

66. A milestone I entry determines whether a new acquisition program is warranted. If approved, costs, schedule, and performance objectives are established.

67. Stephen J. Hadley and Henry Cooper, *Briefing on the Refocused Strategic Defense Inititative*, February 12, 1991, 9.

68. P. A. DeBiaso, "Space-Based Defense," *Comparative Strategy*, vol. 12 no. 1, 1993, 41.

69. DeBiaso, 42.

70. DeBiaso, 42.

71. DeBiaso, 42.

72. DeBiaso, 42.

73. DeBiaso, 43.

74. Stephen J. Hadley and Henry Cooper, *Briefing on the Refocused Strategic Defense Inititative*, February 12, 1991, 15.

75. Hadley, 17.

76. Hadley, 18.

77. Hadley, 18.

78. Baker Spring, "For Strategic Defense: A New Strategy for the New Global Situation," *Heritage Foundation*, April 18, 1991, 5.

79. Spring, 5.

80. Spring, 5.

81. Spring, 6.

82. Spring, 7.

83. Spring, 7.

84. Spring, 7.

Chapter Two

The Origins of Ballistic Missile Defense in the United States and the Soviet Union and the ABM Treaty

This chapter examines the origins of ballistic missile defense in both the United States and the Soviet Union. In particular, each of the United State's and Soviet Union's ballistic missile defense systems architectures will be examined, which include both endoamospheric and exoatmospheric interceptors, that is, a missile designed to target threat ballistic missiles both inside the atmosphere and in space. The ballistic missile defense systems of each country eventually led to a layered ballistic missile defense system that was composed of both an exoatmospheric interceptor and endoatmospheric interceptor.

Initially the United States policy toward ballistic missile defense will be examined up to the signing of the Antiballistic Missile Treaty in 1972. This section deals with the organizational infighting between the Air Force and Army for the role of ballistic missile defense. The Army's initial enthusiasm was a quest for a strategic role in the aftermath of World War II. The section proceeds to examine the many concept designs for ballistic missile defense from the NIKE-ZEUS system through to the SAFEGUARD system. The section concludes with the signing of the ABM Treaty.

The Soviet Union's ballistic missile defense program is examined in turn. The Soviet approach to ballistic missile defense is that it did not require its technology to be proven before the system was deployed. This is proven with the deployment of the ballistic missile defense system around Leningrad. The main Soviet ballistic missile defense program was focused around Moscow. The ballistic missile defense structure around Moscow is examined exhaustively along with its subsequent upgrades after the signing of the ABM Treaty.

The final section examines the ABM Treaty. This outlines the narrow and the broad interpretations of the ABM Treaty. Although the

ABM Treaty was signed in 1972, the narrow and broad debate did not essentially take place until the United States renewed its interest in ballistic missile defenses with the Strategic Defense Initiative in 1983.

The United States Ballistic Missile Defense Prior to the ABM Treaty

The Army's Quest for the Role of Ballistic Missile Defense

The V-2 ballistic missile attacks on London during World War II sparked a quest for defenses against ballistic missiles. The United States in the expectation of a such a threat initiated a research and development program into ballistic missile defense shortly after the end of World War II. The first ballistic missile defense program was born out of the NIKE program which was focused on developing defenses against bombers. This program, named the NIKE II study which was initiated in March 1955, was primarily intended to examine air defense requirements for the 1960s, but intelligence assessments of Soviet ICBM capabilities led to a complementary study on Anti-Ballistic Missile Missiles.[1]

The NIKE II study saw the emergence of a new missile, the NIKE-ZEUS, a three-staged, solid-propellant missile designed to carry a nuclear warhead of 400 pounds. Combined with the missile came the ZEUS system that included advanced radar equipment and communication links to tie the subsystems together.[2] The main elements of the NIKE-ZEUS consisted of various radars and the ZEUS rocket. The radar was composed of the forward acquisition radar (FAR) and local acquisition radars (LAR). The former was a surveillance radar capable of scanning the entire visible sky within a 1,000 mile range. The FAR would acquire the ICBM and maintain continuous surveillance at a range of around 600 miles. It would provide information on the incoming target to the LAR. The LAR would track the target, acquire trajectory data, and assign this data to the target track radars (TTR). The TTR would take over automatically from the LAR and provide continuous and precise trajectory information to the computers, which would determine the intercept point.[3] The nuclear warhead of the ZEUS was designed to explode around 100 feet away from the reentering warhead and destroy it around 75 miles away from the area to be defended.[4]

In 1956 Secretary of Defense Wilson in a memorandum relating to the air defense mission, divided the areas of responsibility and gave the army the role of terminal defense and the air force control over area defense.[5] The army was thus responsible for the development of a missile defense system that could be based near a vital potential target such as a city. The Air Force, concerned about the strategic role which the army hoped to achieve, criticized the ZEUS system that the army was developing for point defense. The air force put forward the argument that the key to deterrence was offensive capability. The air force position was strengthened with the Gaither Report in 1957 which was tasked with examining civil defense and the vulnerability of Strategic Air Command. The report declared that it would be after 1962 before a limited defensive capability against Soviet forces was possible and that deterrence by Strategic Air Command bomber forces was the best defense. However, the dispute was settled by the intervention of the new Secretary of Defense McElroy. He solved the dispute between the army and the air force over ballistic missile defenses by giving the army the primary responsibility for developing the ABM system including the ZEUS missile and the air force work on the radar systems and the command and control systems.

The conflict between the air force and the army over the role of ballistic missile defense continued until the findings of a Scientific Advisory Board (SAB) were made known. The SAB's Nuclear Panel chaired by Dr Edward Teller in October 1957 recommended that the air force should vigorously pursue the development of ballistic missiles and reconnaissance satellites. Concerning ABM systems, the committee recommended that the air force pursue a research and development program.[6] The reports findings led Secretary of Defense McElroy to assign the army the primary responsibility for the ballistic missile defense mission.[7]

From NIKE-ZEUS to SENTINEL to SAFEGUARD

The Kennedy administration under Secretary of Defense McNamara undertook a review of ballistic missile defenses in January 1961. This review assessed the technical feasibility and the cost effectiveness of the ZEUS system. The estimated costs of the ZEUS system of $16 billion to defend a significant portion of the country, along with the technical immaturity of the system, meant that the program was restricted to research and development. In July 1962, a ZEUS missile was tested from the Kwajalein test site. Although the initial test resulted in a technical failure the missile passed within two kilometers of the target reentry vehicle. During the period between June 1962 and November 1963, thirteen similar tests were conducted, from which three were partial successes and nine were complete successes.[8]

Secretary of Defense McNamara, out of concern that the ZEUS system would not be able to counter the projected Soviet threat of the late 1960s and early 1970s, decided against deployment. Instead a restructured program was advocated which saw the adoption of the NIKE-X. The development continued of the ZEUS missile, which became the SPARTAN, but a second interceptor, the short-range, high-acceleration SPRINT was added. This created a layered defense. The ZEUS system would attack the warheads at an altitude of 70 to 100 miles, the SPRINT would then intercept the remaining warheads at an altitude of 20 to 30 miles after the atmosphere had eliminated any of the decoys.[9] The addition of the SPRINT missile meant that low altitude nuclear detonations would occur. This made the NIKE-ZEUS radars vulnerable. The radars were subsequently housed almost completely in hardened concrete structures, with the exception of the exposed flat surfaces which could be made moderately blast resistant.[10]

The NIKE-X replaced the older radar with a new phased array radar. The new phased array radar used an antenna with several fixed faces, each of which had an array of radiating elements. One such antenna could generate several beams of radio pulse and rapidly aim them electronically. The speed and accuracy with which these beams could be targeted meant that one radar could perform several functions and service a number of attacking reentry vehicles and defending missiles.[11] The idea of NIKE-X was to cover the northern approaches to the United States by large phased array radars with large footprints. These would then sort the attacking reentry vehicles according to destination and pass the track details to smaller radar sites which would direct the interceptor missiles.[12]

In 1966 there were several reasons for not deploying the NIKE-X. The high cost of the Vietnam War severely limited the defense budget and hence the availability of funds for ballistic missile defense. The scientific community's skepticism of the capability of the NIKE-X system and its ability to counter decoys added impetus to the decision on nondeployment. Secretary of Defense McNamara's own opposition to ballistic missile defense was in sharp contrast to the Joint Chiefs of Staff who favored beginning work on long-lead time components.

In September 1967 the decision was taken to deploy a ballistic missile defense system against the emerging Chinese threat. Secretary of Defense McNamara, in an attempt to convince the Soviet Union that the system was not aimed at them, renamed the ballistic missile defense system to SENTINEL. The additional reasoning McNamara gave for fielding the system was that the system could be used to protect U.S. Minuteman missile fields and hence enhance the ability to deter a nuclear attack by the Soviet Union.[13] The system consisted of the SPARTAN area defense, and SPRINT terminal defense of twenty-five major cities. This included six of the long-range Perimeter Acquisition Radars (PAR), 17 of the shorter range Missile Site Radar, (MSR), 220 SPARTAN missiles, and 480 SPRINT missiles.[14] In 1968 the army began the process of establishing bases for the deployment of the SENTINEL system.

The incoming Nixon administration conducted a review of the U.S. strategic programs which included ballistic missile defense. President Nixon decided upon program I-69 which concluded that missile defenses would be built at twelve sites depending upon how the strategic situation evolved.[15] The establishment of the sites would be through a phased deployment program that cost $800-900 million in the first year. This level of funding allowed construction to begin at two phase-one: sites air force bases at Malmstrom, Montana, and Grand Forks, North Dakota. This marked a change of emphasis from population defense to silo defense. The remaining ten sites would be subject to annual review by the President's Foreign Intelligence Advisory Board. This decision to defend military bases was announced on 14 March 1969 and the program was renamed SAFEGUARD. In early 1970 the expansion phase of SAFEGUARD was announced with the addition of six sites to the two authorized by Congress. Construction was to begin on one site, Whiteman Air Force Base, Missouri, while preliminary work would begin on five other sites, one of which included Washington, D.C. The SAFEGUARD system consisted of both an area and terminal defense capability. It was comprised of the same components as SENTINEL but deployed them with a reordered set of priorities. Although SENTINEL had the option for ICBM force defense, it was mainly configured to protect cities. SAFEGUARD was to protect two Minuteman sites and later, if required, SAC bomber bases.[16] In contrast to SENTINEL, which moved on a fixed basis, SAFEGUARD deployment was to be adjusted according to need.

The SALT negotiations brought about the ABM Treaty which saw restrictions being placed on ballistic missile defenses. The United States continued with the development of a treaty-compliant ABM system at the Mickelsen SAFEGUARD complex. This was located 100 miles northwest of Grand Forks, North Dakota, and was to defend 150 Minuteman missiles. There were two types of missiles employed in the SAFEGUARD system. The high altitude SPARTAN missile was a three-stage, solid propellant rocket with a nuclear warhead that killed warheads by blast and by X-rays that were lethal to warheads several miles away.[17] The second missile was SPRINT, which was designed to operate in the earth's atmosphere and also carried a nuclear warhead. The two missile systems formed a layered defense. SPARTAN was to attack the incoming "threat cloud" of warheads, boosters, and decoys in space, while SPRINT would attack the surviving warheads that had penetrated the atmosphere. However, on 2 October 1975, one day after SAFEGUARD became operational, Congress voted to deactivate the system after DOD studies had shown that Soviet missiles with multiple warheads would be able to overwhelm the system.

The Reorientation of the Ballistic Missile Defense Program

The ABM Treaty and the decision to dismantle the SAFEGUARD ballistic missile defense system led to a shift in focus from deployment to research and development. The research and development was directed at maintaining the U.S. technological capability as a hedge against a possible Soviet breakout of the ABM Treaty's restrictions. The army was developing a follow-on missile defense system called Site Defense which featured a modified SPRINT interceptor (SPRINT II) which had a greater accuracy, an expanded capacity for maneuvering, and better maintainability.[18] The system also included an improved radar system composed of smaller, less vulnerable radars and a commercially proven computer. Congress however instructed the army to redirect its Site Defense program from a prototype development to research and development.

There were two components to the army's reoriented ballistic missile defense program. The first was an advanced technology program aimed at producing major innovations in missile defense components. The second was the Site Defense project which was turned into a broad systems technology program. Within this environment research and development focused on terminal defense for missile silos rather than area defense. The improved capabilities of infrared sensors combined with high-capacity computers produced the hit-to-kill, or kinetic kill interceptors.[19] Prior to this, the accuracy of guidance systems meant that nuclear warheads were required to assure a reasonable kill probability.

Soviet Ballistic Missile Defenses during the Cold War

The first indication of Soviet research and development into ballistic missile defense appeared with the construction of the Soviet missile defense test site near Sary Shagan which began in 1956. The Sary Shagan site was to be identified later as the center for all Soviet missile defense testing.[20] The location of the site approximately 1000 miles from the ballistic missile range at Kasputin Yar made it an ideal site for testing the interception of long-range missiles. The site was in a remote region of the Soviet Union and was located at a distance in the interior to make U.S. monitoring from the periphery difficult.[21]

In the late 1950s U.S. intelligence reports attested that the Soviets were producing encouraging results. Indeed this is borne out by the establishment in 1958 of an independent V-PRO (Protivoraketnaya Oborona: anti-missile defense forces) component for defense against missiles within the Air Defense Forces.[22] There were tangible signs that the Soviet missile defense program was advancing with the deployment of early warning radars around the periphery of the Soviet Union which were similar to U.S. missile tracking radars. Indeed intelligence obtained from a U-2 in 1960 showed that considerable progress had been made towards the development of a missile defense capability and was continuing.[23]

In 1961 Soviet nuclear missiles launched from Kasputin Yar were detonated at high altitudes over the experimental ABM radar at Sary Shagan. This was to test the nuclear effect on the radar's tracking capability. There were also unconfirmed reports that the test had included the interception of an ICBM by an antiballistic missile fitted with a nuclear warhead.[24] In addition to testing the effects of a nuclear blast over an ABM radar, the nuclear tests were used to develop the high-yield warheads that were required for exoatmospheric interception.

The Soviets began to deploy a system that was suspected as being capable of providing a defense against ballistic missiles. U.S. reconnaissance satellites detected that site preparation similar to the test beds of Sary Shagan were taking place near Leningrad in 1962. The sites were placed across the flight corridors which U.S. ICBMs would fly to reach the western part of the Soviet Union. The configuration of the launch sites resembled that of anti-aircraft surface-to-air missiles, and it is probable that the Leningrad system was an attempt to achieve ballistic missile defense capability through modifications to air defense technology.[25] There was considerable conjecture in the West as to the capabilities of the system against ballistic missiles. These claims were based upon the presumption of the performance of the Griffon missile. However, the Griffon program was technologically not ready at this time. The site was dismantled in 1965 due to the interceptor missile's poor performance and the lack of adequate data processing equipment.

In October 1962 work began on an ABM-type radar near Moscow. In January 1966 Secretary of Defense McNamara linked the Soviet exoatmospheric interceptor missile "Galosh" which had been first displayed in November 1964, with the system around Moscow. The system was initially composed of eight ABM complexes, four each to the east and the west of the capital about forty-five miles from Moscow. However, by 1967 work continued on only six of the complexes and in 1968 four complexes had been abandoned. When the system became operational in 1970 only four were activated and the number of ABM launchers had diminished from 96 to 64.[26] The missile complexes were composed of four engagement radars and sixteen missile launchers in two batteries of eight, with two large battle-management radars for the system.[27] A network of long-range "Hen House" radars deployed near the periphery of the Soviet Union supplied the early warning information. They had a detection range of 6000 kilometers and used "billboard array" antennas: two to scan in azimuth, two in elevation and one in a circular pattern. The Hen House network compensated for the Soviets" lack of forward-based early warning stations.

The battle management was provided by two large "Dog House" and "Cat House" radars. These were A-frame radars with ranges of up to 3000 kilometers and they provided the tracking radar with target acquisition information and assigned targets to both the tracking and the interceptor-guidance radars.[28] The Dog and Cat House radars were phased-array and allowed several beams to be generated at once, and to search the sky in microseconds thus enabling the radar to target several targets simultaneously. Dog House pointed north to track incoming U.S. ICBMs, and Cat House pointed south, which enabled it to cover Chinese ballistic missiles. The battle management radars supplied target acquisition information to the Try-Add engagement radars which controlled the final aspect of the ballistic missile defense system. Each of the complexes contained a set of two identical installations with a larger "Chekhov" target-tracking radar and two smaller radars, one used for tracking known as the "Flat Twin" and the other for guiding the interceptor missiles to their targets, known as the "Pawn Shop." This configuration implied that two missiles would be launched against a single target and tracked by the two radars.[29]

The Galosh interceptor missile was a multistage, solid fuel antiballistic missile with a range of at least 200 miles and fitted with a nuclear warhead in the one-two megaton range. It was intended for exoatmospheric interception and area defense. It was capable of defending not only Moscow but also the northwestern part of the Soviet Union which contained a high proportion of Soviet industrial capacity. Indeed U.S. intelligence in 1966 indicated that the Galosh was capable of producing the X-ray effect, that is, the ability to neutralize an ICBM's guidance equipment and fissionable material at considerable distances from the ABM's detonation.[30] In 1968 a new version of the Galosh was introduced which included a loiter capability. The ABM's bus, once it had reached the apogee of its trajectory, could coast while ground radars discriminated attacking warheads from decoys. Once the target was distinguished the ABM's engine was restarted and the interceptor guided toward the reentry vehicle.[31]

During the 1970s the Soviets did not attempt to deploy the permitted number of ABM launchers under the ABM Treaty; instead they focused on improving their existing system. Between 1972 and 1976 there were fifty-five Soviet ABM tests, including tests of high acceleration missiles utilizing more advanced (inertial and infrared) guidance systems than the Galosh.[32] The period between 1978-1980 saw half of the sixty-four deployed Galosh launchers dismantled, and the above-ground launchers were replaced with underground silos. This reduced the interceptor's vulnerability. The Dog House and Cat House radars were replaced by the faster and more efficient large phased-array radars (LPARs), which were capable of detecting and tracking many objects simultaneously. This network of LPARs included the controversial Krasnoyarsk radar.

This period saw Soviet attempts to upgrade surface-to-air missiles to give them a ballistic missile defense capability. This effort was focused on the SA-5. In 1973 and 1974 SA-5s were tested around fifty times in conjunction with strategic ballistic missile flights.[33] The missile was limited in its capabilities as an ABM in that it had a relatively modest acceleration rate and was vulnerable to saturation by decoys and multiple warheads. In 1981 the U.S. Department of Defense regarded the ABM capabilities of the SA-5 as negligible.[34]

In the early 1980s the single-layer system surrounding the Moscow site was modified to add a second layer of shorter-range but increased acceleration missiles. This would enable the distinction between missile warheads and decoys to become clearer. There were two new interceptor missiles: the SH-04, which had the ability of stopping and starting its engine to assist in decoy discrimination, and the SH-08 or Gazelle, with a very short range (less than 100km) and armed with a low-yield nuclear warhead.[35] The introduction of these two missiles increased the effectiveness of the ballistic missile defense system. There were reports in the mid-1980s that the Soviets had a rapid reload capability for ABM interceptors. Indeed, there were indications that the Soviets had an underground automatic reload system which could double or even triple the number of allowed interceptors under the ABM treaty.[36]

The early 1980s saw further Soviet attempts to upgrade surface-to-air missiles for ballistic missile defense purposes. The SA-10 was thought to be the equivalent of the U.S. Patriot. The SA-10's radars were more advanced than earlier SAM systems, although the interceptor missile was believed to be too slow to be effective against ballistic missiles.[37] In 1987 a mobile version of the SA-10 was deployed. By 1989 the SA-10 system was believed to account for about 15 percent of all

Soviet strategic SAM launchers, and was sited primarily around Moscow.[38] The SA-12 was considered to be the most ABM capable of all the Soviet surface-to-air missiles to date.[39] An improved version along with a mobile capability of the SA-12 "Giant" was deployed by the late 1980s. There has been speculation concerning the possible internetting between the LPAR network and the radars of the Moscow system and the SA-10 and SA-12 systems regarding a nation-wide ballistic missile defense system.[40] Such a system would have served in a strategy of damage limitation.

In 1989 the number of antimissile launchers had been increased to the maximum permitted under the auspices of the ABM Treaty of one hundred. The construction work on the "Pillbox" phased array radar located at Pushkino, north of Moscow, began and was completed in 1990. The radar had a pyramid structure with four faces, each displaying a phased-array radar giving the radar a 360 degree capability. It was reported that the Pushkino radar was four times the size of the U.S. Pave Paws radar. This radar provided overall battle management for the Galosh ABM system and would receive missile tracking data from Soviet early warning systems. It was capable of tracking as many as 1000-2000 targets simultaneously along with the ability to guide the long-range interceptors to their target.[41]

During the 1990s Russia found it difficult to maintain the Moscow ABM system at the level it had reached during the late 1980s. The main difficulty concerned the LPAR radars which experienced problems and three of which were reported as no longer operational.[42] Also the collapse of the Soviet Union meant that its integrated network of radars providing early warning of missile attack were sometimes located in the newly independent republics. These sites continued to function but were subject to each of the state's relations with Russia.

The Anti-Ballistic Missile Treaty 1972

The ABM Treaty was one of the clearest indicators of the mutual agreement on preserving from space, weapon systems and their associated systems by the United States and the Soviet Union during the Cold War period. It is for this reason that the ABM Treaty will be examined as an example of the consensus that existed during the Cold War over the use of military space. The ABM Treaty signed in 1972 permitted 200 interceptors in addition to test and training launchers. As a result of the 1974 Protocol to the ABM Treaty, the number of interceptors permitted was reduced to 100 at one deployed site, with a number of additional launchers at test ranges.

The ABM Treaty was a consequence of arms control preferences coupled with an attempt to stabilize the United States-Soviet relationship. The theoretical underpinning of the treaty was the concept devised in the United States of assured destruction, and what was later to become mutual assured destruction. The theory of assured destruction incorporated the ability to deter an attack by the Soviet Union on the United States by having the capability to inflict in retaliation unacceptable damage on Soviet society. This concept, which was adopted by the United States in the 1960s, was mirror imaged onto the Soviet Union as mutual assured destruction.

The theory assumed that once the Soviet Union had acquired the capability to inflict unacceptable damage on the United States a stable situation of mutual deterrence would follow. This situation once achieved could be cemented by arms control agreements. It therefore followed that it was in both the United States' and the Soviet Union's interests not to threaten the assured destruction capability of the other. Any weapon systems that threatened the other's assured destruction capability should be avoided. It is in this context that a belief grew in the United States that ABM defenses should not be deployed.

The theory argues that an ABM defense would reduce the ability of the adversary to achieve assured destruction. This would then lead the adversary to acquire more offensive capability in order to overcome the effect of the ABM system. This would then lead to offensive arms racing. Hence ABM defenses were thought to be destabilizing and should be avoided. It was this theory that led President Nixon in 1969 to begin strategic arms limitation talks with the Soviet Union that produced a temporary agreement on offensive arms and the ABM Treaty in 1972.

The period between 1972 and 1983 saw little ballistic missile defense activity in the United States. It was not until President Reagan's speech in March 1983 that the issue of the ABM Treaty rose to the fore. The SDI research and development program saw two interpretations of the ABM treaty. The narrow interpretation and the broad interpretation are discussed below.

The Interpretations of the ABM Treaty

The narrow view of the treaty argues that Article V of the treaty forbids development, testing, or deployment of any future ABM systems and components other than those that are fixed land-based systems.[43] The so-called broad interpretation views Article V in the context of Article II and Agreed Statement D, a view that would permit development and testing of systems based on "other physical principles" (i.e., than those specified in Article II), but conditions their deployment on agreement between the parties on specific limitations.

A study by the Legal Advisor of the State Department led to the conclusion that the Treaty language is ambiguous and can be read to support the broad interpretation of the treaty. The three primary provisions demonstrate this:

> •Article II(1) defines an ABM system as "a system to counter strategic ballistic missiles or their elements in flight trajectory, currently consisting of" ABM interceptor missiles, ABM launchers, and ABM radars.
> •Article V(1) provides that the parties agree "not to develop, test, or deploy ABM systems or components which are sea-based, air-based, space-based, or mobile land-based."
> •Agreed Statement D provides as follows:
>
> In order to insure fulfillment of the obligation not to deploy ABM systems and their components except as provided in Article III of the Treaty, the Parties agree that in the event ABM systems based on other physical principles and including components capable of substituting for ABM interceptor missiles, ABM launchers, or ABM radars are created in the future, specific limitations on such systems and their components would be subject to discussion in accordance with Article XIII and agreement in accordance with Article XIV of the Treaty.[44]

The narrow interpretation rests on the meaning of article V(1): it says no deployment of "ABM systems or components" other than those that are fixed land-based. But this does not settle the issue of future systems and components. This rests on the meaning of the term "ABM systems or components." Is this limited to systems or components based on then-current technology, or does it include those based on future technology?

In order to answer this question, the definition of "ABM system" in article II(1) must be examined. Proponents of the narrow view believe that this is interpreted as anything that "could serve the function of countering strategic missiles in flight falls within the definition."[45] They argued that the three components identified in that paragraph—missiles, launchers, and radars—are listed as the elements that an ABM system is "currently consisting of" and that all future components of a system that satisfies the definition are also covered by article II(1). When these definitions are interpreted in this way proponents can rely on article V(1) as a ban on development, testing, and deployment of all nonfixed, land-based systems or components, whether current or future.[46]

The narrow interpretation of the ABM Treaty has some shortcomings. The premise that article II(1) defines "ABM system" as including all future systems and components is difficult to sustain. This provision can be read to mean that the systems contemplated by the treaty are "those that serve the functions described *and* that currently consist of the listed components."[47] The treaty's other provisions consistently use the phrases "ABM system" and "components" in contexts that reflect that the parties were referring to systems and components based on known technology.

For example, Article II(2), describes the "ABM system components listed in paragraph I of this Article," to include those that are being tested, operational or under construction, thus indicating that the definition in article II(1) was intended to describe the actual components covered by the treaty. Also, article V(2) sets limits on the types of "launchers" that may be developed, tested or deployed, reflecting in the same article as the alleged prohibition on future mobile systems and components, concern for one of the current components listed in article II(1).

Agreed Statement D poses a problem for the narrow view of the treaty. Nothing in that statement suggests that it applies only to future systems that are fixed and land-based. It addresses and even presupposes the development of all ABM systems and components that are "based on other physical principles." The narrow view would render this provision superfluous. If article II(1) extended to all ABM systems and components, based on present as well as future technology, then article III implicitly would have banned all future fixed land-based systems and components. These arguments highlight the ambiguities of the ABM Treaty.

Once an agreement has been found to be ambiguous, under international law guidance must be sought on the circumstances surrounding the drafting of the treaty. In the case of the ABM Treaty the negotiating record was consulted to determine what most accurately reflected the parties" intentions. Sofaer reached the conclusion that,

> although the U.S. delegates initially sought to ban development and testing of nonland-based systems or components based on future technology, the Soviets refused to go along, and no such agreement was reached. The Soviets stubbornly resisted U.S. attempts to adopt in the body of the treaty any limits on such systems or components based on future technology; their arguments rested on a professed unwillingness to deal with unknown devices or technology . . . *The parties did not agree to ban development and testing of such systems or components, whether on land or in space.*[48]

The negotiating record contains strong support for the interpretation of "ABM system" and "components" limited to those based on current physical principles. The Soviets specifically sought to prevent broad definitions of these terms, and the U.S. negotiators acceded to their wishes. Although some U.S. negotiators of SALT I talks assert that they achieved a total ban on the development, testing, and deployment of future mobile systems, the record of the negotiations fails to demonstrate that they actually succeeded in achieving their objective. The issue of early warning satellites as components are only a concern when they are used in a ballistic missile defense system. That is, they are free from Treaty restrictions as long as they are not linked up with a missile defense system.

In October 1985, the Reagan administration reiterated a broad interpretation of the Treaty, under which the development and testing of ABM systems and components would be permitted without restraint-principally to clear the way for testing defenses in earth orbit. As Codevilla argues:

> Hence the ABM treaty's treatment of "futuristic" ABM systems is literally the only thing it could be: an agreement to discuss, and to agree about, specific limitations of future weapons as those weapons are created. But, as everyone knows, an agreement to agree is not a deal but an expression of sentiments that each side may regard as it wishes. If it were otherwise, there would be no need for further discussion or agreement.[49]

In late September 1985 a report emerged that averred the development (but not deployment) of nontraditional ABM systems and components (those based on "other physical principles") was not constrained by the Treaty. The Soviet Union had refused to accept such limits during the negotiations.[50]

The Office of the Legal Adviser of the State Department undertook its own review of the negotiating record and confirmed the earlier findings. The State Department review did read the Treaty as banning the deployment of nontraditional ballistic missile defense. The government's Special Arms Control Policy Group convened at the White House. On 4 October, the new interpretation was adopted and it was also decided to offer Moscow five to seven years notice of intent to withdraw from the Treaty.[51]

The Legal Adviser, Abraham Sofaer argued against the narrow interpretation by following three paths: the language of the treaty is ambiguous; the U.S. side tried but failed to obtain Soviet support for banning "exotics"; and the postnegotiation public record is ambiguous as to the U.S. government's policy on the matter. Sofaer's first point is that

Article II(1) of the ABM Treaty is not a functional definition of an ABM system and its components that merely uses traditional components as an example, but is rather a precise definition of what the Treaty is intended to cover. Since only launchers, radars, and interceptors are named, only those things are constrained. The only instance in which the Treaty reaches components based on "new physical principles" is in Agreed Statement (D) associated with Article III (which defines the exceptions to the treaty's overall deployment ban). That Statement, Sofaer noted,

> explicitly allows the "creation" of such systems and components; it requires that limitations on such systems be stipulated only after creation of the systems . . . . Nothing in Agreed Statement D, however, states that it applies only to future systems that are fixed land-based.[52]

Sofaer also concludes that Agreed Statement D was the farthest the Soviets were willing to go on exotics. He argued that "U.S. negotiators persuaded the Soviets to adopt Agreed Statement D by explaining that without it, the Treaty would leave the parties free to deploy future systems or components based on other physical principles."[53]

Finally, Sofaer contends that the U.S. government itself did not adhere unequivocally to the narrow interpretation of the Treaty. He cites a number of statements by U.S. officials regarding the agreement's impact on the development and testing of technologies that do not explicitly distinguish between fixed, land-based technologies and others. He concludes that this reflects at best ambivalence in the U.S. position.[54]

The broad interpretation of the ABM Treaty would permit the development and testing of systems based on other physical principles than those understood in 1972. This would allow the development of the defensive systems such as the space-based interceptor, space-based laser, and the space-based sensor, Brilliant Eyes.

Shortly after the October 1985 endorsement of the broad interpretation by the White House, Richard Perle, Assistant Secretary of Defense, characterized the administration's decision to abide by the narrow interpretation as temporary.[55] In Senate hearings the following spring, Perle described the new version as "the only legal" interpretation of the Treaty and predicted that the administration would see that there was "no rational basis for long-term adherence to the restrictive interpretation."[56]

In July 1986, President Reagan replied to Soviet offers of measures to strengthen the treaty and a pledge not to exercise their right to withdraw for 15-20 years. The letter offered a five to seven year period of treaty observance governed by the broad interpretation, followed by freedom to deploy ballistic missile defense. In October, Reagan placed the same proposal on the table at the Rekjavik summit, lengthening the period of observance to ten years. In turn the Soviet Union proposed that, for a period of ten years, the United States and Soviet Union would "adhere strictly" to the provisions of the agreement.[57] At a news briefing after the summit, the then National Security Adviser, Admiral John Poindexter, confirmed that the administration was offering to delay U.S. withdrawal from the ABM Treaty in exchange for Soviet consent to the broad interpretation of the agreement.[58]

In October 1986, two weeks after the Rekjavik summit, the Reagan Administration clarified its position on "research" and "development" in an address by Paul Nitze. Nitze defined research to include "conceptual design and testing conducted both inside and outside the laboratory."[59] Development "commences with the construction or testing of one or more prototypes of the system or its major components."[60] His definition also tallied with the broad interpretation, claiming allowance for space testing of ABM systems and components based on other physical principles.

The Reagan administration argued that the ABM Treaty, broadly interpreted, allowed the testing of SDI components based on other physical principles in space. The U.S. Congress had imposed a narrow ABM Treaty interpretation on the SDI program—even though the Soviets had never agreed to this interpretation.[61] It was argued by supporters of SDI that this congressional restraint led to suboptimal SDI tests that costed more, took longer, and were more risky than the tests a good engineer would conduct.[62]

At the outset of the Defense and Space Talks, the U.S. position was that the ABM Treaty permits research and experimental work on space-based ABM systems prior to development. "Development" was understood to begin with the field testing of full-scale ABM systems or components, or their prototypes. The U.S. reviewed the negotiating record in the summer and fall of 1985, and concluded that a broad interpretation, which would permit development and testing but not deployment of space-based ABM systems, was fully justified. By September 1987, Moscow had agreed that some testing in space was legitimate under the ABM Treaty, yet the Soviets were unclear as to how much and what kind of testing they viewed the ABM Treaty to permit.[63] At the same time, Moscow had declared that its chief concern was deployment, not testing, implying that it would settle for an interpretation that permit testing so long as the prohibition on deployment continued.[64]

The Democratic majority in the Senate, with the help of a few Republicans, followed Senator Nunn's lead and applied diverse pressures. In early October the Senate voted to withhold funds for any strategic defense testing that would violate the narrow interpretation of the treaty. On November 17, 1987 Congress and the White House reached a compromise over the size of the fiscal 1988 defense budget. The figures leaned slightly toward the administration's position but only because, as part of the understanding, the administration agreed to delay implementing its broad interpretation of the ABM Treaty at least until the following year.[65]

A few days after the December 1987 summit, and with Caspar Weinberger, Richard Perle, Frank Gaffney, and Kenneth Adelman all gone from the administration, Secretary Shultz announced that the administration would no longer insist on the broad interpretation, but would ask Congress to fund SDI testing on a case by case basis. This was a deeper concession than merely announcing a delayed implementation of the broad interpretation; the hope was that through this concession a way could be found around Congress" refusal to fund SDI testing.[66]

Conclusion

The United States and the Soviet Union's approaches to ballistic missile defense differed considerably towards the signing of the ABM Treaty. The United States was initially enthusiastic towards the deployment of a ballistic missile defense system. This enthusiasm however waned shortly before the signing of the ABM Treaty. The proposed SAFEGUARD system containing both exoatmospheric and endoatmospheric interceptors eventually became a pawn in the arms control negotiating process, and with the signing of the ABM Treaty the United States dismantled the SAFEGUARD site. Indeed, the ABM Treaty was the death knell for U.S. ballistic missile defense until the Strategic Defense Initiative in 1983. U.S. domestic politics had made the issue of missile defense sensitive since the coverage was no longer concerned with population defense, but defending missile silos. This was politically impractical for domestic political reasons. The Soviet Union on the other hand continued to be interested in ballistic missile defense both before the ABM Treaty and after its signing. Indeed, considerable work was done on the Moscow ABM site in the period after the ABM Treaty and their interceptors were enhanced along with their associated radar and tracking facilities.

The architectural designs of the United States and the Soviet Union's ballistic missile defense systems were essentially very similar. The two systems were a layered defense with endoatmospheric and exoatmospheric interceptors. The Soviet Union initially focused its efforts principally on exoatmospheric interception for the Moscow ballistic missile defense system. It is not the case that the two sites' similarities were due to mirror imaging, but that it was well recognized that a layered missile defense system offered the best means of protection. The use of the atmosphere in distinguishing between warheads and decoys provided a rationale for the inclusion of an endoatmospheric interception system.

The dismantling of the SAFEGUARD site saw the ballistic missile defense issue more or less disappear from the political scene until the early 1980s. This is in marked contrast with the Soviet Union which continually upgraded and maintained their operational ballistic missile defense system. Indeed, the Moscow site was continually upgraded up to the demise of the Soviet Union in the early 1990s.

The issue of ballistic missile defense was one of the key areas that would have led to the weaponization of space. This would have occurred in the United States with the building of a missile defense system that included an exoatmospheric interceptor. This idea gained considerable political support, but this waned considerably in the early 1970s prior to the signing of the ABM Treaty. However, the Soviet Union built the ballistic missile defense system that included the Galosh exoatmospheric interceptor. This system was maintained throughout the Cold War. The use of space for ballistic missile defense was and remains an extremely important issue for the weaponization of space, and will be addressed again in a later chapter.

Notes

1. Donald R. Baucom, *The Origins of SDI, 1944-1983*, (Lawrence: University Press of Kansas, 1992), 7.
2. Ruth Currie McDaniel, *The U.S. Army Strategic Defense Command: Its History and Role in the Strategic Defense Initiative* 2nd ed. (Huntsville, Alabama: U.S. Army Strategic Defense Command, 1987), 12.
3. Benson D. Adams, *Ballistic Missile Defense* (New York: American Elsevier, 1971), 24.
4. Angelo Codevilla, *While Others Build* (London: Macmillan, 1988), 48.
5. Baucom, 9.
6. Thomas A. Sturm, *The USAF Scientific Advisory Board: Its First Twenty Years, 1944-1964* (Washington, D.C.: Government Printing Office, 1986), 82-83.
7. Baucom, 14.
8. Baucom, 17.
9. Benson D. Adams, *Ballistic Missile Defense* (New York: American Elsevier Publishing Company, 1971), 79-80.
10. Adams, 64.
11. Ernest J. Yanarella, *The Missile Defense Controversy: Strategy, Technology and Politics, 1955-1972* (Lexington: University Press of Kentucky, 1977), 82.
12. Codevilla, 50.

13. Baucom, 37.
14. Kenneth Werrell, *Hitting a Bullet with a Bullet: A History of Ballistic Missile Defense* (Airpower Research Insititute, 2000), 15.
15. Baucom, 41.
16. Benson Adams, 200.
17. Baucom, 92.
18. Baucom, 95.
19. Baucom, 103.
20. Jennifer G. Mathers, *The Russian Nuclear Shield from Stalin to Yeltsin* (London: Macmillan Press, 2000), 11.
21. Sayre Stevens, "The Soviet BMD Program," in Ashton B. Carter and David Schwartz, eds., *Ballistic Missile Defense* (Washington, D.C.: Brookings Institution, 1984), 192.
22. John Prados, *The Soviet Estimate: U.S. Intelligence Analysis and Russian Military Strength* (New York: The Dial Press, 1982), 77.
23. Sayre Stevens, 191-192.
24. Jennifer G. Mathers, 34.
25. Michael J. Deane, *The Role of Strategic Defense in Soviet Strategy* (Miami: University of Miami Press, 1980), 27-28.
26. Jennifer G. Mathers, 61. See also Bruce Parrott, *The Soviet Union and Ballistic Missile Defense*, (Boulder, CO: Westview Press, 1987), 30.
27. Bill Gunston, "Soviet Missiles," in Ray Bonds, ed., *Soviet War Power* (London: Corgi, 1982), 253.
28. Sayre Stevens, 197-198.
29. Sayre Stevens, 198.
30. Jennifer G. Mathers, 62.
31. Jennifer G. Mathers, 63.
32. David Yost, *Soviet Ballistic Missile Defense and the Western Alliance* (Cambridge: Harvard University Press, 1988), 30.
33. Yost, 39.
34. Yost, 40.
35. Yost, 34-5.
36. Yost, 36.
37. Rip Bulkeley and Graham Spinardi, *Space Weapons: Deterrence or Delusion?* (Cambridge: Policy Press, 1986), 145.
38. Department of Defense, *Soviet Military Power 1989* (Washington, D.C.: Government Printing Office, 1989), 51.
39. David Yost, 42.
40. David Yost, 65.
41. David Yost, 37.
42. International Institute for Strategic Studies, *The Military Balance 1998-99* (Oxford: Oxford University Press, 1999), 108.
43. Abraham D. Sofaer, *Statement before the Subcommittee on Arms Control, International Security, and Science* of the House Foreign Affairs Committee, Washington, D.C., on October 22, 1985, 25.
44. Antiballistic Missile Treaty, Agreed Statement D, 1972.
45. Abraham D.Sofaer, *Statement before the Subcommittee on Arms Control, International Security, and Science* of the House Foreign Affairs Committee, Washington, D.C., on October 22, 1985, 25.
46. Sofaer, 25.

47. Sofaer, 26, (emphasis in original).
48. Sofaer, 27, (my emphasis).
49. Angelo Codevilla, *While Others Build* (New York: Free Press, 1988), 183.
50. William J. Durch, "The Future of the ABM Treaty," *Adelphi Papers*, Summer 1987, 22.
51. Durch, 22.
52. Abraham Sofaer, quoted in William J. Durch, "The Future of the ABM Treaty," *Adelphi Papers*, Summer 1987, 23.
53. Abraham Sofaer, "The ABM Treaty and the Strategic Defense Inititiative," *Harvard Law Review*, June 1986, 1974-5.
54. Sofaer, 1980-84.
55. Don Oberdorfer, "Top-Level Fight Led to ABM Policy Shift," *Washington Post*, October 17, 1985.
56. Charles Mohr, "'Option' Sought to Deploy Space Shield Soon," *New York Times*, October 19, 1986, A21.
57. Leslie Gelb, "Reagan Reported to Stay Insistent on "Star Wars" Test," *New York Times*, July 24, 1986, p1; "Excerpts from Speech by Gorbachev about Iceland Meeting," *New York Times*, October 15, 1986, A12.
58. White House, Office of the Press Secretary, *Press Briefing by Admiral John M. Poindexter*, National Security Advisor, October 13, 1996. The U.S. proposal tabled at Rekjavik pledged "strictly to observe" the Treaty provisions for a period of up to ten years "while continuing research, development and testing, which are permitted by the ABM Treaty . . . At the end of the ten year period, either side could deploy defenses if it so chose unless the parties agree otherwise."
59. *Permitted and Prohibited Activities Under the ABM Treaty*, Current Policy No. 886 (Washington, D.C.: Department of State, Bureau of Public Affairs, November 1986.
60. State Department Policy Document.
61. Kim R. Holmes and Baker Spring, eds., *SDI At The Turning Point: Readying Strategic Defenses for the 1990s and Beyond*, (Washington, D.C.: The Heritage Foundation, 1990), 86.
62. Holmes, 86
63. Holmes, 88.
64. Holmes, 88.
65. Holmes, 88.
66. Holmes, 88.

Chapter Three

The Soviet/Russian Approach to Military Space during the Cold War and Beyond

This chapter examines the Soviet military uses of space and Russia's military space activities since the breakup of the Soviet Union. The development of Soviet military space during the Cold War is addressed through the prism of the theories of space power, although these concepts of space power are mainly the product of the United States Air Force's thinking that they provide an insight into Soviet thinking on the issue of the weaponization of space. The Kettering space group headed by Geoffrey Perry was the first to openly examine the telemetry of Soviet satellites and to classify them by missions. The Soviet approach to military space was based on the writings of Sokolovsky which provide an alternative concept to the realm of space. The extent to which this approach is still relevant today will be examined along with the Russian Military Space program and the fate of the military space units in the former Soviet Republics.

The conversion of military equipment for so-called civilian purposes is also addressed. The negotiations with Kazakhstan over the cosmodrome in Baykonur, the only one outside of Russia, are scrutinized. The military utilization of space facilities is analyzed in terms of its contribution to conventional capabilities. Lastly, the cooperation with the United States is examined with Russia both in terms of the potential for cooperation with the global protection system and in the field of launch technology.

The United States and Soviet Union after World War II found themselves at the forefront of a new world. This world expanded out from the surface of the earth into space. The race for supremacy in this new frontier was on, and the Soviets appeared to take the early lead. Today many argue that the race is over, and that Russia cannot continue the legacy in space that has remained since the collapse of the Soviet Union. Indeed one commentator pronounced in relation to the fate of the space assets that, "the Soviets possessed all elements which made up "space power." They exercised these elements, then they lost these elements."[1]

The military was, and still remains at the center of the Soviet space program. The early rockets that fired Sputnik and Yuri Gagarin into space were ICBM derivatives. As part of the military, the space program received vast sums of money, but only portions of the program were publicly open. These programs that were disclosed, along with the general direction of the entire space program, were politically motivated and designed to build the support and admiration of the people while displaying Soviet supremacy to the world. Public support today for space activities is waning. The economic conditions in Russia make it difficult, almost impossible, to justify vast expenditures associated with the space program, and the division of the former Soviet Union also split apart the infrastructure needed to sustain space efforts.

One of the problems with addressing Soviet military space activities is ascertaining the budgetary allocations and priority they were given in the overall defense budget. The figures vary as to what the Soviet Union allocated during the 1980s, although some have indicated between 1 and 2 percent of its Gross National Product out of an overall defense budget of approximately 12 percent of the GNP.[2] A further difficulty with these figures is that much of the space program was part of the military program. Also, some of the space budget, was probably contained in the science budget further adding to the problems for even a rough estimation.

The space infrastructure in the former Soviet Union (FSU) was distributed among many republics; however, the majority was found in only three. Russia had the bulk of the space facilities with around 80 percent of the total capability. Ukraine had around 5 percent of the infrastructure, including the crucial Zenit (SL-16) launch vehicle production facilities and tracking stations. Kazakhstan possessed around 15 percent of the Soviet space infrastructure, including the Bykonur Cosmodrome.[3] This dissection of assets disrupted the supply of essential assets to production facilities, caused disagreements over territorial jurisdiction, and even hindered the actual control of satellites already in orbit.

This chapter examines the development of the Soviet Military Space Program along with the fate of the military space units in the former Soviet Republics. The chapter also addresses the conversion of ICBMs into civilian launch vehicles and illustrates the efforts being made with U.S. cooperation to convert military hardware and capabilities into sources of hard currency. The state of Russia's military space assets since the demise of the Soviet Union is analyzed.

The Development of Soviet Military Space during the Cold War

The 1968 version of the Soviet Military Strategy outlined the Soviet view of the use of space. The military use of space according to the Soviet perspective of space followed three paths. The first was to create space satellite systems to assure combat effectiveness for all branches of the armed services,[4] the second was to prevent other countries utilizing space, and the third was to develop strategic offensive systems to conduct battle in space.[5]

The Soviet space control objectives included the protection of tactical and strategic strike capabilities; support of tactical and strategic operations; protection of client state territories from enemy threats; prevention of the use of space by the enemy for military, political, or economic gain, and unhampered utilization of space assets to further the goals of the Soviet system.[6]

The protection of tactical and strategic strike capabilities from space has been primarily passive in the form of space and ground-based early warning sensors of impending attack. Also, space surveillance of enemy strategic forces or deployments serves this purpose. This was done primarily by the Soviet Union's photoreconnaissance satellites with both land and sea surveillance capabilities.

The use of space to support Soviet tactical and strategic operations was provided by satellites which provided navigational support for troop deployments, resupply, and targeting; command, control, and communications support; weather predictions for planning; reconnaissance for target identification and strike assessment; and intelligence gathering.

The prevention of the use of space by the enemy for military, political, or economic gain was targeted towards NATO. The supply lines and communication links from the United States to Europe depended on satellite support and were a key target for Soviet planners. To accomplish this mission the U.S. space systems would have been attacked, or the ground command and control links might have been targeted.

The formation in 1963 of a special antispace defense establishment called PKO (Protivo Kosmicheskaya Oborona) under the PVO-Strany air defense branch signaled the Soviet Union's intention to seriously develop an ASAT capability. The new unit was given the mission to repel any attack emanating from space.[7]

The Soviet use of launch vehicles of military origin provided it with the option of using tactical or strategic launch facilities instead of normal space complexes in times of crisis. The Soviet ocean surveillance satellites and antisatellites employed a derivative of the SS-9 ICBM and could have been launched from the SS-9 silos across the southern Soviet

Union.[8] The SS-9 boosters which launched the ASATs could be wheeled from the Tyuratam launch site and erected for use in less than ninety minutes.[9] The SS-9 represented the greatest threat to U.S. space systems owing to the response time of the system.

The controversy surrounding the compliance of the Soviet Fractional Orbital Bombardment System (FOBS) system with the Outer Space Treaty was resolved when the U.S. maintained that since a FOBS missile does not remain in space for one complete revolution of the Earth it is not in orbit.[10] Under the proposed SALT II Treaty, Article VII, the eighteen FOBS vehicles, also at Tyuratam, would have been dismantled and under Article IX of the Treaty all future development, testing, and deployment of FOBS would have been banned. Although the SALT II Treaty was never ratified, the Soviet Union did not resume testing which implied that the system was no longer operational.[11]

In the autumn of 1962 the Soviet Union dropped its opposition to satellite reconnaissance.[12] This was in part due to its failure to gain support at the United Nations for such a ban and also around this time the Soviet Union began to use its own photoreconnaissance capability, which began to return photographs in 1962.[13]

The expanding use of space by the Soviet military during the 1970s provided an indirect threat to the United States in that Soviet satellite capability enhanced the Soviet Union's overall war-fighting potential.[14] The cessation of the Soviet satellite interceptor tests in 1971 saw the Soviet Union concentrate on its reconnaissance satellites, in particular the ocean surveillance system capable of tracking U.S. and NATO warships.[15] Indeed the Soviet Armed Forces (the Air Force, the Strategic Rocket Forces, and a specialized ministry level called the "Space Forces") supported space operations by running the launch sites and tracking stations, and by training the cosmonauts.[16]

In the early 1980s the Soviet Union proposed two arms control treaties to prohibit the further militarization of space. These proposals were seen by many as merely political propaganda. The Soviet Union was rapidly expanding its military satellites at this time. Also, these proposals came on the back of the Soviet Union's invasion of Afghanistan and the United States was not prepared to negotiate on arms control measures after this.

A characteristic of the Soviet satellite philosophy of many cheap satellites compared with the U.S. philosophy of a few expensive satellites rewarded it with an inherent replenishment capability.[17] The subsequent requirement of a high launch rate mandated that launch vehicles and satellites be produced in large quantities so that stockpiles were inevitable.

The Soviet Union in 1983-1984 combined its opposition to SDI with a campaign to prevent the testing and deployment of antisatellite weapons; this campaign was abandoned in 1985.[18] This was probably due to the fact the Soviet Union wanted to consider the option of an antisatellite capability against a possible U.S. SDI deployment.

Theories of Military Space Underpinning Soviet Military Space Policy

The Soviet Union's use of space can be categorized as reflecting the broad principle of the sanctuary school of space power. The Soviet deployment of photoreconnaissance satellites along with ocean surveillance satellites follows the space sanctuary philosophy. Hence space should be weapon free and reconnaissance satellites should be used for arms control purposes which strengthens the agreements.

The Soviet Union did, however, develop both the Fractional Orbital Bombardment System (FOBS) and an ASAT capability. These two space systems demonstrated that within Soviet thinking about military space there were views which followed what could be interpreted as the high ground school of space, which views space as the ultimate arena in which to deploy weapons. The FOBS system would fall neatly into this high ground view of space with its ability to strike with weapons of mass destruction in an extremely short flight time. The strong emphasis on an ASAT capability follows the high ground view, but combined with the Soviet military strategy in relation to space control, it would tend to demonstrate the space control school of thought which sees space as another geographical arena from which military operations can be conducted.

The FOBS system following the SALT II negotiations saw the Soviet Union cease testing the system. This action indicated that while the Soviet Union flirted with the high ground and space control schools of thought in relation to military space they moved away from embracing these philosophies fully. The subsequent actions in relation to military space demonstrated that the sanctuary school of space was once again in the ascendancy.

Chapter Three

Global Protection against Limited Strikes: Russian Cooperation

President Bush, in a nationally-televised speech in September 1991, called upon the leadership of the Soviet Union to take immediate concrete steps to permit the deployment of defenses against ballistic missiles. Less than two weeks later, President Mikhail Gorbachev declared that the Soviet Union was ready to consider proposals for nonnuclear defense against ballistic missiles. There was a break in the process as the Soviet Union dissolved. This came about owing to the August Coup attempt in 1991 and its subsequent failure with the struggle between the reformers and those in favor of the status quo. This process resulted in the emergence of Boris Yeltsin as President of Russia.

In October 1991 statements in support of joint U.S.-Russian missile defense were being made by some Soviet military and political officials, suggesting that the Soviet Union was reconsidering its previous opposition to cooperative missile defense. Indeed, Velikhov, when asked about opponents of joint BMD, reportedly stated, "there are practically none among either designers or the military. The critics of this proposal in both Russia and the United States are, rather, maniacs obsessed with old ideas and they have no influence."[19]

The new government of Russia recognized the threat of ballistic missile proliferation and the subsequent need for enhanced defenses. Boris Yeltsin announced in January 1992, in a speech to the United Nations Security Council, that he was "ready to work out and subsequently create and jointly operate a global system of defense."[20] Yeltsin subsequently discussed the issue with President George Bush at Camp David in February 1992, and endorsed a joint U.S.-Russian Statement on a Global Protection System (GPS) at the Washington summit meeting in June 1992. At the summit meeting a press release was issued which included the following:

> The two Presidents agreed it was necessary to start work without delay to develop the concept of GPS. For this purpose they agreed to establish a high-level group to explore on a priority basis the following practical steps:
>
> The potential for sharing of early-warning information through the establishment of an early warning center;
> The potential for cooperation with participating states in developing ballistic missile defense capabilities and technologies;
> The development of a legal basis for cooperation, including new treaties and agreements and possible changes to existing treaties and agreements necessary to implement a Global Protection System.[21]

The first meeting of the high-level group was held in Moscow 13-14 July 1992. It was headed by Assistant to the President Dennis Ross on the U.S. side and Deputy Foreign Minister Georgiy Mamedov on the Russian side. The group agreed to establish three working groups: one to develop thinking about the GPS concept itself, a second group to explore areas for possible technology cooperation, and a third to explore common efforts on nonproliferation. The senior group would retain responsibility for legal issues associated with a GPS.

The second meeting of the high-level group was held in Washington 21-22 September 1992. Four topics were discussed: (1) technology cooperation, (2) nonproliferation activities, (3) further elaboration of the GPS concept, and (4) further discussion about issues associated with the legal basis for GPS.

It is interesting to note the following quote related to Yevgeni Velikhov:

> At a recent conference in Erice, Sicily, with scientists from a number of nations around the world, Yevgeni Velikhov was there wearing an SDI tie. He is vice president of the Russian Academy of Sciences, and directs many Russian institutes involved in SDI-like work, with which we might cooperate. In the past Velikhov has written papers vehemently opposed to our program. He and Andrei Kokoshin were co-authors on a number of such papers. *At the conference, Velikhov said he thought we had to move away from mutual assured destruction, as the basis for our planning and strategic relationship, towards mutual assured protection.*[22]

The U.S.-Russian high-level group established by the two presidents met in July and September 1992. That group was informally known as the Ross-Mamedov Group. The two delegations established working groups to deal with the overall GPS concept, with technical cooperation, and with nonproliferation. The United States were also discussing the legal basis for GPS.[23]

Baker Spring, a policy analyst with the Heritage Foundation, quoted in the Washington Times, said that the proposed U.S.-Russian warning system could be used in a strategic defense command and control system—one of the long lead-time components of ballistic missile defenses.[24] One of the benefits of cooperation was that it might give greater warning time to intercept launches.

The Russian leadership apparently at the time shifted its security focus from the Western threat to the threat from the South. According to Sergei Rogov, deputy director of the U.S.A. and Canada Institute, a primary threat to Russian security is the threat of "certain former republics

[that] may find themselves under the influence of regional power centers like China or Iran."[25] Some Russian military officers argued at the time (in statements to American audiences) that missile defense has become important because the leaders of Third World states that are acquiring a missile capability may not be deterred by the threat of retaliation. As Senator Dole apparently believed at the time when he was the Republican presidential nominee:

> If the will in Russia is still there, if the details can be worked out, we can go forward together. U.S.-Russian cooperation on missile defense could be a key element in a global effort to combat nuclear terrorism and missile blackmail.[26]

In the Yeltsin period there were three prevalent views in Russia on the ABM Treaty: work within the treaty (Deputy Defense minister Kokoshin favored this approach); change it to allow GPALS; repudiate it. Dr. Savelyev's[27] view is that "the treaty is a symbol of past relations and could become a bomb to explode the development of new relations. It had its role, whether positive or negative, but now it should be eliminated, 'scrapped.'"[28] At an International Security Council meeting where this view was aired, when asked how widespread these views were in the Russian government, and how much opposition existed to them, General Batenin (who was Chief of the personnel staff of Vice President Rutskoi and Counselor to the Foreign Minister) and Dr. Shlykov (who was Deputy Chairman of the State Committee on Defense) responded "that there are some who oppose change, and it may take time, but the views just expressed are more and more widespread."[29]

General Samoilov, (Counselor for Military Affairs to President Yeltsin), said that the ABM Treaty was not inviolable, and that there were different interpretations of its provision.[30] He reported that a linguistic analysis had been performed on the text of the treaty, which had led to the following questions:

> 1) The U.S. and Russian texts of the treaty are not the same—there are wide and deep philosophical differences reflected by the differences.
> 2) Supporters of the ABM Treaty hold positions that are not literally viable
> *Now, he said, they are not politically or strategically viable.*[31]

Dr. Savalyev argued that a ballistic defense system is an important deterrent to proliferation. If the United States and Russia had cooperative strategic defense, it would be doubtful that a third government would go against them. Although there was no single answer to the threat of proliferation, Dr. Savelyev reiterated that the greatest nuclear threat is that posed by ballistic missiles, not 'suitcase bombs.'[32]

Nevertheless, there remained points of contention on the subject of missile defense and cooperation with the United States. In particular, the question of space-basing of interceptors, the feasibility of significant outlays of capital for defenses in a time of economic crisis, the possible threat to the Russian nuclear deterrent posed by defenses, and fears of a one-way technology transfer from Russia to the United States at "fire sale" prices. These concerns were primarily voiced within the military, and by members of the old Soviet foreign policy and arms-control establishment.

The U.S. approach to Yeltsin's concept of a Global Protection System was to embrace it gingerly. This occurred for three reasons: first, GPS was seen as a possibility of changing the thinking about ballistic missile defense. The ABM Treaty resulted in missile defense being thought of as something bad and destabilizing. The focus on a GPS system was thought about as a possible vehicle to change that thinking.[33]

Second, it was thought that by focusing on the GPS concept it could help change thinking in Russia. President Yeltsin's call for cooperation on a GPS system was a breakthrough in the attitude of the former Soviet Union on these issues.[34] Cooperation in a GPS system could provide a context in which Russia could accept the deployment of U.S. defenses against ballistic missiles and the changes in the ABM Treaty required to allow those deployments.

Third, the focus on a GPS system could help change thinking in the United States. If Russia was ready to work on defenses against limited ballistic missile attacks, then the most skeptical critics in the United States would have to give way. It would help provide relief from the ABM Treaty and the constraints it placed on deploying defenses.

Participants in the GPS would establish and operate a global protection center, which would carry out certain tasks. They would share information on sources of proliferation and the uses of proliferated technology. They would register pre-launch notifications of launches of ballistic missiles and space vehicles. They would share specified information of all launches of missiles detected by national sensors—such as time of launch, location of launch, number of missiles launched, direction of flight, and the like. They would assist one another to develop their own national technical means of warning and defense against limited ballistic missile attacks. Center participants could undertake planning activities, engage in exercises, and develop models to support cooperative defensive operations against such attacks. The center would be a forum in which individual states could work out cooperative agreements by which the assets of one nation might be used to defend the territory of another against limited ballistic missile attack. At the same time, participants would retain control of the national assets they committed to a Global Protection System.

In early 1993 and 1994 the initial treaty modifications discussions with the Russians were unproductive. The Russians quickly understood that the United States had given them a veto under any future planned improvements in TMD. In a poor example of negotiating style, the United States revealed its intentions to seek "clarifications" in the treaty so that the proposed interceptor systems would be allowed to attain speeds of up to 5 kilometers per second. At this speed these systems could have significant capabilities against 'strategic" missiles. The 5-kilometer-per-second requirement was necessary in order to engage such medium range missiles as China's CSS-2 that travels at about 4.5 kilometers per second.[35] The Russians understood the 'strategic" implications and refused to accept the proposed changes.

The Clinton administration's position at the time on the ABM Treaty can be seen from an article in the Washington Times:

> [Clinton's] statement says the ABM Treaty is the "cornerstone" of strategic stability; that regional systems can be built if they don"t violate or circumvent the treaty; that regional systems can be deployed if they don"t pose a "realistic" threat to strategic forces; that regional defenses will not be limited "in number and geographic scope" consistent with short-range missiles.[36]

Administration efforts to negotiate with Moscow a distinction between defenses against long-range strategic missiles and short-range theater missiles drew Republican concern that the administration may be willing to accept too many limits on development of theater defenses, particularly on the speed of interceptors.[37]

Yeltsin's turnabout has enormous implications for the U.S. debate on missile defense. The standard arguments against defenses—which were all based on the assumption of Soviet hostility—were now without foundation.[38] For the past decade, every major argument against SDI has been based on the premise that the Soviets would oppose U.S. deployment of strategic defenses. Critics argued that to advance with SDI would violate the 1972 ABM Treaty. They contended that the Soviet Union would never agree to revise the treaty for the purpose of expanding defenses. During a brief period during the Yeltsin period Russian attitude had changed. However, the lack of enthusiasm towards ballistic missile defense during the Clinton administration meant that the GPALS concept was effectively ended.

The Fate of the Military Space Units since the Breakup of the Soviet Union

There has been much speculation about Soviet military capabilities in space since the World War II. The following quote highlights Soviet Military Space Doctrine:

> The Soviet Armed Forces shall be provided with all resources necessary to attain and maintain military superiority in outer space sufficient both to deny the use of outer space to other states and to assure maximum space-based military support for Soviet offensive and defensive combat operations on land, at sea, in air, and in outer space.[39]

This assessment was compiled in 1984, prior to the changes in the former Soviet Union. A failure of the Soviet economic situation which can be evidenced today is the lack of foresight in harvesting the technological advances inside the space industry. Aside from a few instruments for their own use the Soviets were rarely able to devise a practical space-related benefit from the Soviet industrial base.

Space units are not like other military units, and have been affected differently by military reforms. Space units of the former Soviet Union were assigned to launch and support the functioning in orbit of spacecraft for scientific, national economic, and military purposes, manned spacecraft, and orbital stations. Theirs was a special role in that they had to simultaneously perform military and national economic tasks.[40] With the collapse of the Soviet Union, the republics were eager to confirm their sovereignty and control over equipment on their territory.

Training and manning levels are another area that separates space units from other military units. Space units operate some of the most advanced technology found in Russia. Training personnel to operate this equipment properly is often time consuming and the necessary skills are frequently difficult to acquire. A reduction in military personnel of approximately fifty percent will hit space units particularly hard. Shortages of manpower were already being felt by 1992. At some facilities,

> the composition of the duty shifts operating costly spacecraft in flight is just one-half of what it should be. Only the greatest professionalism and responsibility by the personnel of the command and control centers, their correct understanding of the difficulties which the entire nation is experiencing, can explain the virtual absence of unsuccessful sessions for controlling the spacecraft.[41]

Almost half of the command and control facilities for multipurpose spacecraft are located outside Russia. These centers could not easily be replaced. Some officials advocated moving all command and control operations to Russian territory, but this required significant expenditures. In fact Russian privatization policies were used to help ease the financial constraints being felt. The financing of a new generation of launch vehicles was hoped to be achieved by issuing shares in Russia's space agency.[42]

Soviet Cosmodromes

The former Soviet space program utilized three cosmodromes to launch spacecraft: Kapustin Yar, Plesetsk, and Baykonur. Kapustin Yar was the site of the first Soviet launch in 1947, but now handles only occasional missions. A total of eighty-three launches from Kapustin Yar had occurred through the end of 1991, however only one of these was after 1986. Many of the missions launched from the cosmodrome were shifted to Plesetsk. Plesetsk was the busiest of all the former Soviet cosmodromes, it saw a total of 1366 vehicles launched through 1991.[43] Its northern location makes it the better site from which to place satellites into polar and highly elliptical orbits. Both Kapustin Yar and Plesetsk are located within Russia, therefore the ownership of the sites was not disputed.

Baykonur is the only cosmodrome outside Russia, and Russian officials felt that the loss of Baykonur would have been a serious blow to Russia's space program. Baykonur's location further south takes advantage of the earth's rotational energy that assists efforts to place satellites into orbit. This geographical location factor allows for the use of heavier payloads or less powerful launch vehicles for missions. Baykonur is the only site capable of manned launches. Some Russians called for further development of Plesetsk to take over some of the missions now restricted to Baykonur, however officials estimate it would take more than ten billion roubles to convert Plesetsk into a Russian Baykonur. During an April visit to Plesetsk, President Yeltsin stated that Russia could not make this kind of economic investment at the present time, but that gradually more missions would be transferred to Plesetsk. Yeltsin went on to acknowledge some of the problems experienced between Russia and Kazakhstan by saying, "our Kazakhstan friends are a little capricious . . . agreement between Russia and Kazakhstan on Baykonur is needed."[44]

Russian President Yeltsin sought a ninety-nine-year lease on the 600 square mile sites, but Kazakhstan sought a shorter commitment because it wants to operate the site once it is technically and economically capable. The deal which was struck gives Moscow a twenty year lease on the cosmodrome, but includes a clause allowing the lease to be extended by ten years.[45]

Condition of Launch Facilities

The confusion over the ownership of Baykonur and the financial contribution each party must make has not allowed the construction, repair, and the maintenance of facilities to run smoothly. "The main problem is still the failure to implement interstate agreements on Baykonur which were concluded on 25 May 1992 with the intention of settling the sides" mutual claims."[46] The problems were especially apparent at the Buran launch complex.

Energia Science and Production Association General Designer Yu P. Semenov rejected the idea of mothballing the complex, and asserted that Buran was scheduled to be repaired for its second test flight within a year. Other specialists were optimistic. According to cosmodrome employees, ." . . repair and restoration work on the Buran launch complex alone would take a year. This was largely the result of the systematic pilfering of equipment from unguarded launch pads. Everything that can be stolen, has been stolen. . . . components made out of copper and other metals are misappropriated by the kilogram daily."[47] Experts also claimed that the landing strip constructed for the shuttle no longer had the capability for an automatic landing by an unmanned space plane, as was performed in 1988.

Even with these funding problems, the cosmodrome continued to launch operations. In 1992, twenty-three of the forty-eight satellites launched by Russia were put into orbit from Baykonur. Work continued on the Zenit launch pads, and three or four new Zenit launches were expected. Finally, preparations were in progress to accommodate launching modified SS-35 ICBMs for commercial use as part of the "START" system.[48] The Start system was basically to use Soviet ICBMs which were to be eliminated under the conditions set out under the Strategic Arms Reduction Talks (START) and use them as boosters to put satellites into orbit.

The decrees on 7 May 1992 that formally established Russia's armed forces acknowledged the Military Space Forces along with the other branches of the military. The extent to which the Russian military was

Space-based Missiles

Space-based weapons would be extremely powerful. Supercompact nuclear weapons suitable for arming miniaturized missiles could be carried onboard orbital platforms or launched from space. The existence of a program to develop such a weapon was revealed in by Korotkevich, a scientist and adviser to Yeltsin. In the early 1980s Korotkevich said, when he was an aide to Oleg Baklanov (minister of ground machine building) he was in charge of developing a new generation of ground-, sea-, and space-based missiles.[53]

The Soviets accused the U.S. of planning to deploy tens of nuclear missiles aboard the shuttle and also made it clear that they could do the same, on the Buran. A Soviet plan to deploy such miniaturized nuclear missiles on board the Buran would explain the cryptic remark made by the designer of the Buran program, that the Buran was conceived as a counter move to the perceived threat that the U.S. shuttle might make a pass over Moscow on its first orbit, carrying a dangerous payload, and bomb Moscow to smithereens.

The deployment of nuclear weapons does not violate any existing disarmament treaties. According to Avduesvskii, 'suborbital and partially orbital flights through space by ICBMs and other objects with nuclear weapons and other types of weapons of mass destruction on board, are not prohibited by the 1967 Outer Space Treaty, since ICBMs and these other objects do not belong to the category of "objects, launched into orbit or deployed in space."[54]

Apparently Russia has already began work on developing miniaturized missiles suitable for deployment in space.[55] These were known as third-generation nuclear weapons, in which a doubling of yield is achieved with a hundredfold reduction in weight compared with existing nuclear weapons.

Advantages of Space-based Weapons

The short time of warning associated with an attack from space is a noticeable advantage of space-based weapons. Space-based offensive missiles deployed in an orbit at an altitude of 500km (300 miles), could reach ground targets in only one minute. Basing nuclear missiles in space would make them largely invulnerable to boost-phase and midcourse interception.

It would appear that Russia has adopted a different approach to the realm of military space from that adopted by the West. Russia appears to be developing its military along the lines of utilizing space as a further arena in which to conduct war. The concepts of space-based nuclear weapons appear to validate this statement and would if deployed cause a potential enemy great problems. Although the West believed that it was the first to think of using the arena of space for missile defense it appears that the former Soviet Union was indeed thinking about such activities, long before President Reagan announced the Strategic Defense Initiative. The Revolution in Military Affairs in which space plays a pivotal role appears to be defined in a different way in Russia. Russia appears to be using the realm of space not in a revolutionary way, but as a continuation of military activities.

Until 1994 Russia was able to maintain its military space systems at Cold War levels.[56] It had been anticipated that the former Soviet military space program's capabilities would not be maintained.[57] However, by the end of 1992 the Russian space program began maintaining and even expanding the former Soviet satellite capabilities. The increase in satellite launches continued into 1993 with the launch of twelve new military spacecraft during the first four months of the year, including three advanced navigation satellites, two electronic intelligence satellites, two imaging reconnaissance spacecraft, two missile warning satellites, two communication satellites, and a new-generation ocean surveillance spacecraft.[58] In 1994 there were twenty-six dedicated military launches of which included seven photoreconnaissance satellites, two ELINT satellites, one EORSAT, two early warning satellites, eight communication satellites and two low-altitude photoreconnaissance satellites.[59]

Russia had not only managed to maintain the majority of its 1990 satellite capabilities, by 1995 it had in some areas expanded its capabilities. In particular, while the number of ELINT 3 satellites had dropped from six in 1990 to three in 1995 this was a phase-out which was being replaced by the ELINT 4.[60] The ELINT 4 constellation in 1995 remained at a full complement of four spacecraft. A similar pattern was seen with photoreconnaissance satellites; although launch numbers were drastically reduced from the days of the Soviet Union when a figure approaching thirty-five were launched each year, this was due to a newer type of reconnaissance satellite. The newer reconnaissance satellites were capable of lasting much longer than their Soviet counterparts.[61] Russia launched far fewer navigation, early warning, and ocean surveillance satellites than the Soviet Union did, but this was because their satellites were lasting longer. Consequently in 1995 two of the three navigation networks were maintained at four and six satellites, while another was slightly expanded to twenty-one; one early warning system was maintained at nine satellites while the geosynchronous launch detection network was expanded from three to four.[62]

The prioritization of military space objectives during the mid-1990s is made even more remarkable given the problems the Russian aerospace community has faced since the dissolution of the Soviet Union. Indeed these efforts were a continuation of those under the Soviet Union in the late 1980s in which it is estimated that the Soviet space program was 85-90 percent military and about 70 percent of Soviet space launches were military.[63]

U.S. and Russian Cooperation of Space

Conversion: Alternative Use for Military Equipment

The START treaties permit two methods of disposing of the ICBMs over the allowed limits. The first is cutting up the missiles into small sections. The second is converting the missiles into launch vehicles to be used only for peaceful purposes. A project with the name "START I" proposed the use of SS-20 and SS-25 missile technology for civilian purposes. "START-I" is being developed by the "Kompleks" scientific and technical center, with financial backing from the IVK commercial joint-stock company. The developers cited several tenets as the basis of their work on the project:

> Objective conditions established in the services market; ensuring maximum continuity of scientific and production activities of enterprises and, consequently, preserving their intellectual and production potentials; accomplishing tasks with a maximum return in minimum time periods by using amassed experience.[64]

IVK looks for the project to be a source of hard currency. Another selling point of the project is that Russia needs to reduce the number of missiles that have already been produced and this enables it to maintain the industrial base which manufactured these missiles. S. Zinchenko, vice president of IVK, claims that the "START-I" program has already saved approximately 5,000 jobs and involved over 15,000 personnel (10,000 of which are servicemen from strategic rocket forces subunits).[65] In the future, IVK looks for "START-I" to be a strong competitor in the space launch market.

Due to advances in technology, many satellites in the future will be much smaller than those launched today. The "START-I" launch vehicle is designed to carry these smaller satellites into low earth orbit at a fraction of the cost of today's commercial vehicles. Yuriy Solomonov, director of "Kompleks," stated that launch services using the "START" system will cost potential customers $7-$10 million, compared with $60-$80 million paid by companies today.[66]

The "START" system has other potential advantages for Russia. Officials cite the "high reliability demonstrated when . . . destroying SS-20 missiles by the launch method. All 72 launches were successful, and the figure is even more impressive when combined with previous launch statistics for these missiles—approximately 250 trouble-free launches."[67] Another advantage of the system is its transportability. Everything required for a launch is self-contained: aiming system, power supply, and the necessary mechanical and transporter erector equipment. COCOM restrictions on technology used in many satellites have hindered Russia's attempts to enter the commercial space launch market. The "START" system's transportability allows it to be delivered wherever a potential customer prefers. This may allow Russia to sidestep COCOM restrictions by conducting launches outside Russian territory, a definite advantage from the Russian point of view.[68]

This system illustrates the efforts Russia is making in conversion of military capabilities into sources of hard currency. The first test launch of the "START-I" occurred in March 1992 from Baykonur. Baykonur, however experienced problems associated with the dispute over control of the cosmodrome between Russia and Kazakhstan, which has subsequently been resolved.

Lockheed Martin has been marketing Russian and Ukrainian launch services, whether from Russia or through innovative arrangements for launch elsewhere. Lockheed is the company most deeply involved, through the LKE International (Lockheed-Khrunichev-Energia) joint venture, but several others, including Boeing, are attempting to develop prospects involving Ukrainian launch vehicles and a variety of converted Russian missiles. The Lockheed-Khrunichev-Energia joint venture (LKE International) is marketing Proton launch services internationally, for both geostationary and low-Earth orbit satellites.

NASA purchased $650 million in goods and services from Russia during fiscal years 1994-1997, by far the largest transfer of U.S. public funds to the Russian government and private organizations. Such purchases entail some political risk in the Unites States, as well as risk to the space station if the Russian government and enterprises are not able to perform. Some U.S. observers question the wisdom of supporting part of the Russian aerospace industry, which provided much of the technological substance for the Soviet threat to the United States.

The SDIO actually initiated the first major private sector imports of Russian space technology beginning in late 1990, when it sought to import Topaz 2 space nuclear reactor hardware and space thrusters. SDIO used private firms as its purchasing agents for these procurements.

The United States must decide how much of its industrial base should be maintained to meet national security needs and to ensure access to space. Making use of existing Russian technology could reduce the amount of research and development required of U.S. companies, resulting in reduced costs, but it could undercut the development of U.S. capabilities in certain areas. Because the space industry is indispensable to the security of the United States, many argue that the United States should develop and maintain its own capabilities in certain critical areas to prevent any weakening in its own technological base. The situation by 1998 was such that foreign investment into Russia was approaching U.S. $800 million per year,[69] twice as much as Russia itself allocated to its budgeted for military space activities. Apparently a similar amount is budgeted for military space activities.

The Current States of Russia's Space Assets

On 15 June 1998 six Russian military Strela 'store-and-forward" communications spacecraft were launched from Plesetsk Cosmodrome aboard the SL-14 Cyclone booster.[70] The satellites were designated Cosmos 2,352-2,357. In addition to this, Russia launched two new military imaging reconnaissance satellites on 24 June and 25 June 1998 from Plesetsk and Baikonur cosmodromes respectively.[71] The first of these was Cosmos 2,358, a high-resolution fourth-generation spacecraft with two film return pods. The second was Cosmos 2,359, a medium-resolution, search and find digital-imaging spacecraft. The two spacecraft operate in tandem with the broad area surveillance satellite used to assist the higher resolution satellite to pinpoint targets. An early warning satellite Cosmos 2361 was launched on the 30 September 1998 and placed into an elliptical orbit.[72] The Glonass program, the navigation satellite network similar to the U.S. GPS system requires twenty-four satellites; however, Russia had only fifteen in the constellation. Three Glonass satellites were launched on, 30 December 1998 to build the system up to eighteen satellites.[73]

The launch of three military space missions involving a total of eight satellites indicates that the Russian military space force has a viable capability. This has been degrading steadily since the collapse of the Soviet Union. Russia in the late 1990s was consistently launching around fifteen satellites per year, which were mostly Cosmos designated intelligence gathering, early warning, and tactical communications satellites. This was down by more than 80 percent from the high point in the 1980s and early 1990s.[74] In particular the new reconnaissance satellites ended a three month hiatus when no reconnaissance satellites were in orbit at all. The performance of the Russian space command is regarded as being critical to the success of commercial missions, with international customers observing with interest how well the Russian military conducts the surge in launches.

The aviation and defense sectors were placed under the command of the Russian Space Agency (RSA) in 1999. In 1998 the RSA generated nearly $1 billion in exports for launch services and various other space goods and services.[75] The RSA was renamed the Russian Aerospace and Space Agency (RASA). The renaming was part of a wider process to restructure and encourage the agency to be more competitive on the world market and bring in foreign partners. The other Western nations" space programs that cooperate with RASA include the European Aeronautic Defense and Space Consortium. The space forces in Russia launch reconnaissance satellites for the Russian military as well as launches for commercial purposes as a sideline providing lucrative hard currency for Russia's monetarily constrained space industry.[76]

The importance with which Russia views military space was reiterated with the reinstatement in June 2001 of the Russian Military Space Forces as an independent subservice, which was reported directly to the General Staff.[77] The Russian military space forces were formed in 1992 until the decision was taken in 1997 to place space forces under the control of the Strategic Rocket Forces. This decision to place forces under the Strategic Forces led to a decline in Russian military space capabilities, which the reorganization hopes to halt. In 2001, Russia's satellite constellation was composed of around 100 vehicles of which a significant proportion are close to the end of their operating lives. President Vladimir Putin appointed Anatoly Perminov as commander of the space forces on 28 March 2001. Anatoly Perminov's background and experience in space science in Plesetsk where he commanded over 100 launches, is a further indication of Russian desire to reinvigorate military space activities. Russia has also recently deployed an optical tracking facility in Tajikistan in order to be able to identify space objects up to 40,000 kilometers (25,000 miles) from earth.[78]

Russian Military Space Satellites

Type of Satellite	Name	Dates Launched
Communication	Cosmos 2337-2339	2/14/1997
	Cosmos 2352-2357	6/16/1997
	Cosmos 2384-2386	12/27/01
	Molniya M3-45	8/4/1993
	Molniya M3-46	8/23/1994
	Molniya M3-47	8/9/1995
	Molniya M3-48	10/24/1996
	Molniya M3-49	7/1/1998
	Molniya M3-50	7/8/1999
	Molniya M3-51	7/20/01
	Molniya M3-52	10/25/01
	Geyser	7/5/00
	Globus 1	8/27/00
	Strela 3 (series 96)	12/28/00
	Luch	12/16/94
	Luch I	10/11/1995
	Cosmos 2085	7/18/1990
	Cosmos 2172	11/22/1991
	Cosmos 2291	9/21/1994
	Cosmos 2319	8/30/1995
	Cosmos 2371	7/4/00
	Cosmos 2372	8/28/00
Early Warning	Cosmos 2340	4/9/1997
	Cosmos 2342	5/14/1997
	Cosmos 2351	5/8/1998
	Cosmos 2368	12/28/1999
	Cosmos 2350	4/28/1998
	Cosmos 2379	8/25/01
	Cosmos 2361	9/30/98
Earth Observation	Okeon-01	7/17/99
Navigation	Glonass 50-M10, 51-M11, 52-M12	12/30/98 (all)
	Glonass 51-M12, 54-M13, 54-M13, 55-M14	10/13/00 (all)
	Glonass 80-82	12/30/1998
	Glonass 83-85	10/13/00
	Glonass 86-88	12/1/01
Reconnaissance	Cosmos 2366	8/18/99

	Cosmos 2367	8/18/99
	Cosmos 2369 (Tselina-2)	2/3/00
	Cosmos 2370, 2383	5/3/00, 5/29/01
	Cosmos 2377	5/29/01
	Cosmos 2344	6/6/1997
Electronic Intelligence	Cosmos 2221	11/24/1992
	Cosmos 2228	12/25/1992
	Cosmos 2242	4/16/1992
	Cosmos 2219	11/17/1992
	Cosmos 2227	12/25/1992
	Cosmos 2237	3/26/1993
	Cosmos 2263	9/16/1993
	Cosmos 2278	4/23/1994
	Cosmos 2297	11/24/1994
	Cosmos 2322	10/31/1995
	Cosmos 2333	9/4/1996
	Cosmos 2360	7/28/1998
Electronic Ocean Surveillance	Cosmos 2367	12/26/1999
	Cosmos 2383	12/20/01

Source: *Aviation Source Book,* Aviation Week and Space Technology, January 14, 2002, 171-173 and *SIPRI Yearbook 2002* (Stockholm International Peace Research Institute: Oxford University Press 2002), 660-663.

The table above highlights that Russia continues to operate a military space program, although not on the scale that was witnessed during the Soviet period. This is a function of lifetime and possibly economic considerations. The focus of Russia's current military space program is on maintaining its navigation positional satellite system, the Glonass system, and its satellite communications system. Russia is also maintaining its intelligence satellites both in terms of electronic and electronic ocean surveillance.

Conclusion

The Soviet Union used the realm of space for military purposes. They developed a military space program which incorporated satellites for reconnaissance, electronic intelligence, electronic ocean surveillance, communications, early warning, and meteorology. After the end of the

Cold War Russia initially struggled to maintain its military space capabilities. However, Russia continued its military space program although to a lesser degree than the former Soviet Union. One of the methods Russia used to create investment into its space program was converting ICBMs into launch vehicles and selling this service successfully on the international market. Russia under the guise of the Global Protection System devised by Yeltsin showed interest in the early to mid 1990s in cooperating in an international missile defense system. This system would have included space-based weapons. This shows to a certain degree some acceptance of the weaponizing of space. However, the GPS system was no longer under consideration by 1996 and Russian attitudes to missile defense hardened. As the previous table shows Russia continues to have a significant military space program.

Notes

1. James E. Oberg, Space Power Theory (Colorado Springs AFB: USAFSPCOM, 1998, 51.
2. Christopher Lee, *War in Space* (London: Hamish Hamilton, 1986), 165. These figures are derived from the CIA's open source estimates.
3. Oleg Velikoredchanin, "Cosmonaut Solovyev Defends Accomplishments of Space Program," *JPRS Report Science and Technology—Central Eurasia: Space*, JPRS-USP-92-004-L, 20 July 1992, 12.
4. V. D. Sokolovsky, *Soviet Military Strategy*, ed. Harriet Fast Scott (New York: Crane, Russak, 1975 3rd ed.), 84-85.
5. Sokolovsky, 85.
6. Nicholas L. Johnson, *Soviet Military Strategy in Space*, (London: Jane's Publishing Company, 1987), 198.
7. Johnson, 138.
8. Johnson, 51.
9. Clarence A. Robinson, Jr, "Antisatellite Weaponry and Possible Defense Technologies against Killer Satellites," in Uri Ra'anan and Robert L. Pfaltzgraff, Jr., eds., *International Security Dimensions of Space* (Hamden, CT: Archon Books, 1984), 71.
10. Nicholas Johnson, 135.
11. Johnson, 136.
12. Paul B. Stares, "U.S. and Soviet Military Space Programs: A Comparative Assessment," in Kenneth Luongo and W. Thomas Wander, *The Search for Security in Space* (New York: Cornell University Press, 1989), 28.
13. Stares, 28.
14. Paul B. Stares, *The Militarization of Space, U.S. Policy, 1945-1984* (New York: Cornell University Press, 1985), 135-136.
15. Stares, 140.
16. Oberg, 51.
17. Johnson, 51.

18. Stephen Shenfield, "The Militarization of Space through Soviet Eyes," in Stephen Kirby et al, *The Militarization of Space* (Sussex: LynneRienner, 1987), 138.

19. Velikhov quoted in K. B. Payne, L. Vlahos, and W. Stanley, "Yeltsin's Global Shield: Russia Recasts the SDI Debate," *Policy Review*, no. 62, Fall 1992, 79.

20. President Boris Yeltsin, speech to the U.N., quoted in Graham, D. R., "Missile Defense Capability," *Comparative Strategy*, vol. 12, no. 1, 1993, 39.

21. The White House Office of the Press Secretary, *Joint Statement on a Global Protection System*, June 17th, 1992.

22. Velikhov quoted in H. F. Cooper, "Unsteady Evolution of the Emerging Consensus on SDI," (emphasis added), *Comparative Strategy*, vol. 12, no. 1, 1993, 30.

23. Ronald F. Lehrman II, "Changing Realities," *Comparative Strategy*, vol. 12, no. 1, 1993, 49.

24. Bill Gertz "Star Wars Backers Hail Defense Project with Russia," *Washington Times*, February 21, 1992, 3.

25. Sergei Rogov, quoted in K. B. Payne, L. Vlahos and W. Stanley, "Yeltsin's Global Shield: Russia Recasts the SDI Debate," *Policy Review*, no. 62, Fall 1992, 80.

26. Senator Robert Dole, "U.S.-Russia Should Build Joint Missile Defense," *USA Today*, May 9, 1995, 11.

27. Dr. Savelyev is Vice President of the Moscow Institute for National Security and Strategic Studies.

28. Dr. Savelyev quoted in W. R. Van Cleave, et al., "C.I.S. and Nuclear Weapons: Liabilities, Risks, Proliferation and Strategic Defense," International Security Council Conference, *Global Affairs*, vol. 8, no. 1, Winter 1993, 184.

29. Quoted in W. R. Van Cleave, et al., "C.I.S. and Nuclear Weapons: Liabilities, Risks, Proliferation and Strategic Defense," International Security Council Conference, *Global Affairs*, vol. 8, no. 1, Winter 1993, 184.

30. General Samoilov quoted in W. R. Van Cleave, et al., "C.I.S. and Nuclear Weapons: Liabilities, Risks, Proliferation and Strategic Defense," International Security Council Conference, *Global Affairs*, vol. 8, no. 1, Winter 1993, 185.

31. W. R. Van Cleave, et al., "C.I.S. and Nuclear Weapons: Liabilities, Risks, Proliferation and Strategic Defense," International Security Council Conference, emphasis not in original, *Global Affairs*, vol. 8, no. 1, Winter 1993, 185.

32. Dr. Savelyev quoted in W. R. Van Cleave, et al., "C.I.S. and Nuclear Weapons: Liabilities, Risks, Proliferation and Strategic Defense," International Security Council Conference, *Global Affairs*, vol. 8, no. 1, Winter 1993, 198.

33. Stephen Hadley, "Global Protection System: Concept and Process," *Comparative Strategy*, vol. 12, no. 1, 1993, 4.

34. Hadley, 4.

35. Hadley, 4.

36. Bill Gertz, "Clinton, Yeltsin Agree on Missiles," *Washington Times*, May 11, 1995, 20.

37. Bradley Graham, "Congress to Push For a National Missile Defense," *Washington Post*, September 5, 1995, 1.

38. K. B. Payne, L. Vlahos, and W. Stanley, "Yeltsin's Global Shield: Russia Recasts the SDI Debate," *Policy Review*, No. 62, Fall 1992, 78.

39. Defense Intelligence Agency, *Soviet Military Space Doctrine* (Washington, D.C.: United States Government Printing Office, 1984), vii.

40. Col. Igor G. Makhalov, "Space Units Press Officer Dissents," *JPRS Report Science and Technology—Central Eurasia: Military Affairs*, 16 September 1992, 13.

41. Makahalov, 13.

42. "Space for hire," *The Times* (London: UK), April 14, 1994.
43. Interavia, 531.
44. Tatyana Malkina, "Yeltsin Favors Agreement With Kazakhstan," (text), Moscow *NEZAVISIMAYA GAZETA* (30 April 1992), 1. Translation by JPRS, *JPRS Report Science and Technology—Central Eurasia: Space*, JPRS-USP-92-004, 10 June 1992, 72.
45. "Yeltsin signs lease on Baikonur," *The Financial Times*, (London: UK), March 29, 1994.
46. Anatoly Zakl, "Political, Economic Woes of Baykonur," (text), Moscow *NEZAVISIMAYA GAZETA*, (9 February 1993), 6. Translation by JPRS, *JPRS Report Science and Technology—Central Eurasia: Space*, JPRS-USP-93-001, 25 March 1993, 45.
47. Zakl, 45.
48. Zakl, 45.
49. "Program Describes Military Space Forces" (text), LD2001105793 Moscow Russian Television Network, 2056 GMT (19 January 1993). Translation by JPRS, *JPRS Report Science and Technology—Central Eurasia: Military Affairs*, JPRS-UMA-32-004, 3 February 1993, 48.
50. Hung Nguyen, "Russia's Continuing Work on Space Forces," *Orbis*, Summer 1994.
51. Nguyen, 413-423.
52. Craig Covault, "Mir Fires Beams at Swedish Satellite," *Aviation Week and Space Technology*, April 4, 1994, 71.
53. See Hung Nguyen, "Russia's Continuing Work on Space Forces," *Orbis*, Summer 1994, 413-423.
54. Nguyen.
55. Nguyen.
56. Matthew J. von Bencke, *The Politics of Space: A History of U.S.-Soviet/Russian Competition and Cooperation in Space* (Boulder, CO: Westview, 1997), 190.
57. Craig Covault, "Russian Military Space Program Maintains Aggressive Pace," *Aviation Week and Space Technology*, May 3, 1993, 61.
58. Covault, 61.
59. Bencke, 195-196.
60. Bencke, 191.
61. Bencke, 191.
62. Bencke, 181.
63. Bencke, 192.
64. Yu. Solomonov, "Missile Conversion for Light-Satellite Launch," *JPRS Report Science and Technology—Central Eurasia: Military Affairs*, JPRS-UMA-93-003, 26 January 1993, 34.
65. "'Start' Project Official Claim 'Waiting Line' for Launch Services," (text), Moscow Teleraiokompaniya Ostankino, Television First Program Network, 2000 GMT (26 May 1992).
66. O. Volkov and D. Molchanov, "'Start' Rocket Seen as Key to Small Satellite Market," (text), Moscow *Komsomolskaya Pravda* (5 June 1992), 4. Translation by JPRS. *JPRS Report Science and Technology—Central Eurasia: Space*, (JPRS-USP-92-005), 21 August 1992, 44-45.
67. Solomonov, 36.
68. Vadim Mikhnevich, "'Start-I' Will Orbit Communications Satellite in First Launch," (text), Moscow *DELOVOY MIR* (4 June 1992), 3. Translation by JPRS. *JPRS*

Report Science and Technology—Central Eurasia: Space, JPRS-USP-002-005, 21 August 1992, 46.

69. Oberg, 59.

70. Craig Covault, "Promise and Peril Mark Russian Launch Surge," *Aviation Week and Space Technology*, July 13, 1998, 78.

71. Covault, 78.

72. Marco A. Caceres, "Satcom Market Buffeted by Economic Uncertainties," *Aviation Week and Space Technology*, January 11, 1999, 153.

73. Craig Covault, "Commercial Proton, Soyuz Launch Surge Readied," *Aviation Week and Space Technology*, February 8, 1999, 68.

74. Marco A. Caceres, "Satcom Market Buffeted By Economic Uncertainties," *Aviation Week and Space Technology*, January 11, 1999, 144.

75. Michael A. Taverna, "Russian Aerospace Wants to Come in From Cold," *Aviation Week and Space Technology*, August 23, 1999, http://www.awstonline.com

76. "Russia Launches U.S. Satellite Into Orbit," *SpaceDaily*, http://www.spacedaily.com/news/020822111626.ie1qzb10.html, August 22, 2002.

77. "Russia Turns Military Ambitions to Space," *STRATFOR*, May 21, 2001, www.stratfor.com.

78. "Russia opens space 'window' in Tajikistan," *SpaceDaily*, July 19, 2002, www.spacedaily.com/news/020719080734.rk4cuey0.html.

Chapter Four

China's Military Space Program

This chapter analyzes China's military space program. It traces its origins and identifies Tsien Hsue-shen as its father. It follows on from this, China's space launch endeavors were based upon its ballistic missile rocket technology. Once China had mastered space launch it proceeded to develop its satellite applications with particular reference to the military sphere. Much of China's space programs are deemed to be civilian, but have dual use, especially with regard to military capabilities. These capabilities are outlined and traced and include photoreconnaissance, communications, meteorology, navigation and electronic intelligence. The launch sites and their geographical parameter with regard to military space are analyzed. The political forces behind the space program are outlined, along with the organizations that are involved in the process.

China's distinctive path to space development includes both military applications and international cooperation. Countries involved with the Chinese program include Russia and the former Soviet republics, the United States, Canada, United Kingdom, Germany and Brazil. It is a very important characteristic of China's military space development that it relies heavily on its cooperation with other countries. In the United States this cooperation caused a great political dispute and led to a Congressional investigation, especially in connection with space launch market cooperation.

Toward the end of the chapter, China's interest in weaponizing space is enunciated, in particular, China's interest in developing antisatellite weapons. It has a number of programs to this end, and appears to be seriously considering an active capability to use as a leveler against a potential adversary's space assets.

The Origins of China's Space Program

The person who contributed most to the development of space in China was Tsien Hsue-shen. In 1935 he pursued a scholarship in aeronautical engineering in the United States which he began at the Massachusetts Institute of Technology (MIT) but later transferred after a year to the California Institute of Technology (Cal Tech), where he was under the tutelage of the mathematician Theodore von Kármán. Tsien's interest in rocketry began when he was invited by five fellow students and associates to join a group interested in amateur rocketry. He was the mathematics adviser to the group, and authored his first work on rocketry in 1937, "The effect of angle of divergence of nozzle on the thrust of a rocket motor; ideal cycle of a rocket motor; ideal efficiency and ideal thrust; calculation of chamber temperature with disassociation."[1] The first experiments of the group were presented to the Institute of Aeronautical Sciences and were soon sponsored by the military which saw the potential in their work in terms of aircraft propulsion and ballistic missiles. Indeed in a period of five years funding from the military rose from $1,000 to $650,000.

In 1942 Tsien was active on small rocket motors to help to get aircraft airborne and subsequently assisted in drawing up plans for a missile program. In 1943 he became Assistant Professor of Aeronautics and was one of the cofounders of the Jet Propulsion Laboratory (JPL) from which the U.S. unmanned exploration of the Moon was later guided. The following year he became the first head of research analysis at JPL and by 1945 was working for Kármán in the Pentagon advising on how to harness the advances made in aeronautics and rocketry for postwar defense forces. In May 1945 he was given the rank of temporary colonel of the United States Air Force and was part of the team sent to Germany to assess the Nazi achievement in rocketry and survey their rocket factories and test sites. Tsien also met with the leading German rocket engineer Wernher von Braun who played a fundamental part in the United States' rocket program.

Tsien returned to JPL and published his wartime technical work in a book called "Jet Propulsion."[2] In 1946 he worked at MIT for two years, during which, in 1947, he returned to China for a brief visit where he married, and subsequently in 1950 he became the Robert Goddard Professor of Jet Propulsion at Cal Tech. It was during his tenure at Cal Tech that he gave a presentation to the American Rocket Society in which he outlined his concept of a transcontinental rocketliner capable of flying 400km above the Earth. The following year he predicted that astronauts would travel to the Moon within thirty years. However, in 1952 at the height of McCarthyism, Tsien was accused of being a communist. His security clearances were revoked and he was put in jail and subsequently held under house arrest. For many months whilst the different factions of the U.S. government battled over whether he should be released, put back in jail, or deported Tsien kept up his work on rocket guidance and on how computers could steer rockets during their ascent through the atmosphere.[3]

In September 1955 in an agreement between the United States and Chinese governments Tsien and ninety-three fellow scientists returned to China in exchange for seventy-six U.S. prisoners of war taken in Korea. It was no coincidence that the origins of China's missile program can be dated to 1956, the year after Tsien's return. Although Tsien brought with him the latest theories of rocketry from the United States, China in the early 1950s had just emerged from a long period of turbulence and destruction: the war with Japan, the civil war, and the communist revolution of 1949. China, unlike the United States, had no aircraft factories, test sites, wind tunnels, or the types of facilities Tsien was familiar with in California.

In January 1956 Chairman Mao Zedong proposed the rapid development of science and technology in China to attempt to catch up with the advanced levels in science and economics in the rest of the world. The Supreme State Conference set up a Scientific Planning Commission under the leadership of Prime Minister Zhou Enlai. In February, Tsien presented a report to the Central Committee entitled "Opinion on establishing China's national defense aeronautics industry."[4] After wide consultations with scientists and experts the committee drew up a plan for the "Long-range planning essentials for scientific development, 1956-67" which established fifty-seven priority tasks to ensure China's independence in rocket and jet technology in twelve years.[5] In April 1956, Zhou Enlai presided over the Central Committee Military Commission which invited Tsien Hsue-shen to outline the potential of guided missiles and rockets. Deputy premier Nie Rongzhen was made director of the newly formed State Aeronautics Industry Commission and Tsien Hsue-shen one its members. It issued its first report on 10 May entitled, "Preliminary views on establishing China's missile research." The report was subsequently accepted by the Central Military Commission on May 26 and the administrative machinery to implement its findings was ordered to be set in place.[6]

On 8 October 1956 the Fifth Academy of the Ministry of National Defense was established by the Central Committee of the Communist Party of China to develop the space effort. Within this institution the Rocket Research Institute was established under vice-premier Nie Rongzhen, and its first director was Tsien. However, the leadership in

China realized that the institute needed outside assistance and turned to the Soviet Union. The Soviet Union initially sold to the Chinese the R-1 missile in October 1956 which was not the latest Soviet technology at the time, which had reached the R-5.[7] Nie Rogzhen led a delegation to Moscow in July 1957 in attempt to persuade the Soviet Union to provide greater assistance. An agreement called the "New Defense Technical Accord 1957-87" was ratified on 15 October 1957 in which the Soviet Union agreed to supply missile models, technical documents, designs and specialists.[8] In January 1958 several R-2s were supplied to China along with a hundred Soviet specialists with over 10,000 blueprints and technical documents. Also, fifty Chinese graduates went to Moscow to study. Despite this assistance, the Chinese felt that the Russians were not providing as much assistance as they might have done.

The Soviet-Chinese agreement came to an end in August 1960, with the Sino-Soviet split. This affected the Chinese space program with the returning to the Soviet Union of 1,400 technical advisers, taking their blueprints back with them and shredding the remainder. Along with this was the cancellation of more than 200 joint projects. The split came just before China launched its first rocket on 5 November 1960 which had been made in China. This Chinese version of the R-2 was named Dong Feng 1, or "East Wind" 1.

The development of longer-range missiles took precedence over the development of an earth satellite. However with the successful testing of Dong Feng 2 on June 1964, which had a range of 1500km, Dong Feng 3, which was operational by 1969 with a range of 10,000km and Dong Feng 4, which was tested on 30 January the notion of an earth satellite was restored.[9] A boost to the development of a satellite came when the Academy of Sciences in Beijing established a committee to investigate the desirability of a satellite. In August 1965 the Central Committee and Prime Minister Zhou Enlai approved the project and gave it the project code 651.

The project saw the development of a new rocket Chang Zheng 1 (CZ-1) or Long March 1. The Long March 1 was in essence a civilian version of the Dong Feng 4 missile. The specifications of the rocket were for a liquid propellant rocket able to send 200kg into low earth orbit. However, the Dong Feng 4 did not have enough thrust to put a satellite into orbit. For this purpose a third stage with a solid rocket motor was added to enable the satellite to reach orbit. All the components were brought together in July 1969, and was eventually shipped to Jiuquan on 26 March 1970. The rocket was assembled on the launch pad on 17 April 1970 and was cleared for flight on 24 April. The rocket was successfully launched and with it China became the fifth country to send a spacecraft into orbit, along with achieving the heaviest first launch by any country.[10] A second satellite was orbited on 3 March 1971 and was named Shi Jian I, which was primarily of scientific importance.

The nature of the technology and the overlapping characteristics of the Chinese military and civil space program permitted the parallel development of a missile program and a space launch vehicle.[11] The technology involved in developing the thrust required to place satellites into orbit was very similar to that required for launching ballistic missiles. China was not alone in this regard; both the Soviet Union and the United States developed their launch capability from derivatives of ballistic missiles.

There was a period of four years between the launch of the Shi Jian I and the next satellite. The Ji Shu Shiyan Weixing series of satellites known as Project 701 consisted of three successful launches and one failure during the period between 1973 and 1976. The lack of information regarding this program suggests that it was a military program. It may have been a project to develop satellites for electronic intelligence gathering purposes. The official announcement during the initial launch of the series stated that the satellite was part of "preparations for war."[12] The official history of the satellite program refers to the precise nature of the orbit and that small errors were not acceptable. This was a similar characteristic of the Soviet electronic intelligence satellites, so it could be inferred from this that the Chinese series had a similar purpose. On 26 July 1975 the Ji Shu Shiyan Weixing 1 entered orbit. It was subsequently destroyed after fifty days. The final two launches of Ju Shi Weixing 2 and 3 were launched on 16 December 1975 and 30 August 1976 respectively.

The Fanhui Shi Weixing, Project 911 was the recoverable satellite program. China became the third country to recover a satellite from Earth orbit. The Fahui Shi Weixing program was modeled on the U.S. Discoverer program which was a military reconnaissance program, and it is highly likely that the Chinese used the program for similar purposes. The first launch of the satellite in this program on 5 November 1974 proved disastrous. However the second launch was successfully carried out on 26 November 1975. There were two further launches of this carried out on 7 December 1976 and January 1978. An improved version of this reconnaissance satellite program was initiated after an interlude of four years. The improved version included CCD cameras to test the feasibility of transmitting data in real time. The first launch of this kind occurred in September 1982.[13] China had developed the charged-coupled device (CCD), "push broom scanner" at the Xian Institute of Radio Technology. The Chinese were also using the CCD cameras to develop a remotely piloted vehicle battlefield surveillance system.[14] One or more of China's previous low-altitude satellites had used a different imaging system to relay pictures to Earth by radio signal. This system was believed to be funded by the Chinese military. The ten-foot in diameter tracking dish for receiving this imagery was developed at the Xian Institute of Radio Technology.

The first Chinese communication satellite program was called Project 331 and began in April 1975. The problem of achieving a geostationary orbit from a launch site with a high latitude (Jiuquan is 41.1 degrees north) was somewhat ameliorated with the creation of a new site. The new site was located in southern China at Xi Chang at 28.25 degrees north, which is at a similar latitude to Cape Canaveral. To support this mission, the PRC increased its tracking, telemetry, and control (TTandC) network which included three ships,[15] along with the Xichang command and control center and a network control center at Weinan, which is near Xian.[16] The network was interconnected via satellite to enhance its capabilities. The complete network is composed of eight ground tracking stations, of which three were mobile.[17] A large scale C^3I was built in the Western Hills, a northwest suburb of Beijing. The center was satellite-linked and strengthened with counter-electronic warfare measures. By the mid-1980s China had established over 2000 stations to receive satellite communications.[18] China benefited in the area of C^3I capabilities from the U.S.-Soviet Cold War rivalry, with the United States supplying U.S. electronic warfare systems and personnel assistance from the CIA's Office of SIGINT Operations, and training for PLA operators.[19] Two receiver stations were built along the Sino-Soviet border.

The communications satellites were termed the Dong Fang Hong series. The first attempt was unsuccessful with the third stage of the rocket not firing, and subsequently not achieving the correct orbit. However, the second mission on 8 April 1984 was successful and the satellite was called Shiyan Tongbu Tongxin Weixing (experimental geostationary communications satellite). The satellite established China's first 200 satellite-based telephone lines and connected Beijing with Urumqi, Lhasa, Hohhot, Chengdu, and Guangzhou.[20] Three strategic coaxial cables for long-distance telecommunications were laid. One extends from Beijing to Hangzhou which connects Nanjing and Shanghai, the other Beijing to Guangzhou, and the final one from Chengdu to Shanghai. The second cable in particular represents a significant improvement in China's communications network. To handle sensitive information, the open wires have been gradually phased out with underground and underwater cables.[21] In addition to this in the 1980s China developed a satellite communications signals intelligence capability for monitoring international satellite communications, along with an associated deception capability.[22]

China has not used the realm of space for early warning of missile attack purposes. Instead they have relied on China's space tracking network which contains large phased-array radars.[23] This work started in the 1970s. The network also tracks space targets. One of the radars is positioned on a mountain slope at an elevation of 1,600 meters, near Xuanhua.[24] The radar was commissioned into service in 1976. The principal radar was the 7010 radar, which allocated targets to other tracking sensors, the 110 radars.

Chinese Launch Vehicles

The Long March 1 is a three-stage version of the ballistic missile Dong Feng 4. The Long March 1 made the first two launches in the Chinese Space Program. However the Long March 2 that was capable of launching recoverable satellites superseded the Long March 1. The Long March 2A, the first of the series, made only one failed attempt to launch a recoverable satellite. The new launcher, Long March 2C, flew successfully fourteen times during the period from 1975 to 1993.[25] The Long March 2C has a longer second stage and is fitted with a smart dispenser that enables small communication satellites to be placed into orbit. The LM-2C is the linchpin of the Chinese commercial launch market.[26] The Long March 2D in 1992 was introduced to place heavier payloads into orbit. The launch capability of the 2D is 3,400kg into low orbit. The Long March 2E was designed to place satellites in a geostationary orbit. It has two stages and is able to place 9 tons into low orbit. To achieve this the Long March 2 was fitted with four liquid-fuel strap-on boosters. This vehicle is able to carry a solid-rocket booster to send a satellite into a geostationary orbit. The LM-2E provides a substantial lift capability to low Earth orbit at a reasonable cost.[27]

The Long March 3 was created to provide the capability to launch 1,400kg to geostationary orbit. It first came into service in January 1984. The Long March 3A is able to launch twice the weight of the Long March 3 into geostationary orbit. The Long March 3B is the most powerful Chinese rocket equivalent to Russia's Proton. Essentially the Long March 3B adds the four strap-on boosters to the Long March 3A to enable a payload of 4.8 tons to achieve geostationary orbit.[28]

The Long March 4 was developed in the 1980s in order to fly meteorological satellites into polar orbits. The Long March 4 is an enlarged version of the Long March 2 but it is fitted with a new third stage. The Long March 4 has a more powerful restartable engine, and is able to lift 4.2 tons into low orbit or 2.8 tons to polar orbit.

The rocket engines in China were given the designator YF (yeti fadong—liquid type engine). There are a core of four rocket engines, of which there are a number of variants. The four principal types are the YF-1 to YF-3 series, the YF-20 to YF-22 series, the YF-40 series, and the YF-73 and YF-75 series.[29] These four types have been modified to be used in the range of the Long March launch vehicles.

The YF-1 was a liquid-fueled rocket engine that had a thrust of 28 tons. The Long March 1 used a combination of four YF-1s in order to provide a thrust of over 100 tons. This combination was called the YF-1A. The YF-3 was the other early rocket motor developed originally as part of the second stage of the Dong Feng 4. The YF-3 operated at an altitude of 60 kilometers. The YF-3 was used for the Long March 1 with an additional solid fuel upper stage added.[30]

The YF-20 engine along with later modified versions were used for the Long March 2, 3, and 4 launch vehicles. It was first introduced in 1975 on the Long March 2 as part of a cluster of four engines to provide 280 tons of thrust on takeoff. When part of the cluster formation it is designated YF-21. The YF-22 engine was designed to ignite at altitude and is subsequently used for second-stage rockets. It was first used on the Long March 2 in 1975 and is a variant on the YF-20.

The YF-40 series was developed as a third-stage engine for the Long March 4 rocket which was introduced in 1988. As a third-stage engine it has a smaller thrust than the first and second stages, but has a longer burn time of around 320 seconds.

The YF-73 and FY-75 engines are liquid hydrogen fueled and are third-stage restartable engines. They are used in order to put large communication satellites into geostationary orbit and have to be restartable, firing to enter Earth orbit, and refiring around fifty minutes later to transfer it to geostationary orbit.[31] In mastering this technique China became the next country behind the United States, Soviet Union, and European nations consortium. The YF-75 was an improved engine which provided a thrust of 8 tons, compared with YF-73 which had a thrust of 4.5 tons, and was used in 1994 on the Long March 3A.[32]

Political Forces behind the Space Program

The politicians whose actions played a critical role in the space program were Zhou En Lai, Lin Biao, and the Gang of Four.[33] Zhou's interest came about since he believed that China's accomplishments in space would be translated into international prestige. Lin Biao's interest in space mirrored Zhou's interest in that he believed it would afford China greater prestige on the international stage. The Gang of Four's interest was based on their attempt to move the focus of space operations from Beijing to Shanghai. They had established the Shanghai Space Research and Production Base there. Some of the work, most notably the Long March program, was duplicating that was undertaken in Beijing. This was done since the Gang of Four had parochial interests in Shanghai.

The military space program under Deng Xiaoping focused on geosynchronous communication satellites.[34] These satellites combined with earth imaging satellites were vital for command, control, and intelligence. In 1975 the development of the communications satellites were included into the State Plan. The first geosynchronous communication satellite was launched on 8 April 1984 aboard the Long March 3 booster. Later in 1986 space was accorded the highest priority status in the technological program.

There were two policies which led the Ministry of Astronautics Industry (subsequently merged into the Ministry of Aerospace Industry) to commercialize its products. First, was China's economic reform policy that reflected a call from Deng Xiaoping in March 1978 for defense industry sectors to devote their efforts to economic growth.[35] It is of note to mention that among the defense industry sector, the space industry is a major force. Second, was a defense industry reform policy in 1983 that was reflected in a systematic transformation of defense-oriented industries into civilian ones.[36] A major consequence of these two policies was that the budget provided by central government was dramatically reduced. In 1987 the budget was 0.035% of GNP compared with the United States 0.52% and the Soviet Union 1.5%.[37] This austere economic environment led to the Long March vehicle and satellite service being placed on the international market to generate foreign currency.

Chinese Launch Sites

China has three launch sites. The oldest launch site in China and most frequently used site is the Jiuquan Space Center which is in the Gansu Province and is located in Shuang Cheng-tzu (41.2°N, 100.1°E).[38] The second launch site is Xichang Satellite Launch Center and is in Sichuan Province, and was built in 1984 to launch the Long March 3 series of satellite launchers. This is the site China uses to launch its geostationary satellites above the equator since it provides maximum advantage of the Earth's rotational momentum. It is located in the southeastern part of China at (28°N, 103°E).[39] In the adjoining valley a launch pad for the CZ2-4L and the CZ3-4L, 1,000 meters from the original launch pad was built. This allowed launch capacity to double from six to twelve launches per year.[40] The third site was built to accommodate the Long March 4 series of satellite launchers in 1988 at Taiyuan (38°N, 112°E).[41] This site is used for launching satellites which require higher inclinations and polar orbits such as weather and communication satellites.

Organizations Involved in Chinese Space Policy

The Commission of Science, Technology, and Industry for National Defense oversees space activities in China.[42] There are however a number of organizations that report to it; these include the China National Space Administration (CNSA) and the China Aerospace Corporation that operates as a private company. The component of the China Aerospace Corporation that is responsible for marketing commercial space launch services is the China Great Wall Industries Corporation. The Chinese Academy of Space Technology has overall responsibility for the development and production of spacecraft. The China Satellite Launch and Tracking Telemetry and Control General organization is responsible for the operation of the launch sites and provides tracking and telemetry services.

The Chinese space industry is combined together within the China Aerospace Corporation (CASC) which was founded in June 1993. The primary organizations within CASC are: the Chinese Academy of Launch Vehicle Technology (CALT), which is composed of thirteen institutes and seven factories; the Shanghai Academy of Space Technology (SAST), also known as the Shanghai Bureau of Astronautics (SBA) composed of seventeen institutes and eleven factories; and the Chinese Academy of Space Technology (CAST) that has fourteen institutes.[43] CALT and SAST both compete for launch vehicle programs, although CAST specializes in satellites.

Current PRC Space Forces

The focus on economic development while improving military technologies is particularly evident in the space sector. The current satellites are dominated by dual use systems, such as meteorological, communications, and remote-sensing satellites.[44] Equally important is China's policy of acquiring technology through cooperation with other countries, along with direct purchase. This has been done since it is hard to distinguish civilian and military space systems. The relative success of China in this regard means that it is likely that it will continue to do this in the future in order to broaden its development of its space systems. This will have corresponding benefits to the military space sector. The uses China would put its space forces to in the event of a war would almost definitely include satellite imagery to monitor defense developments in Taiwan and other supporting countries. Indeed there is evidence to suggest that the PRC does this already with its reconnaissance satellite.[45]

The Chinese seem to be developing capabilities to counter overhead satellite reconnaissance. They have developed through practice a method with which a division combines hiding, lying low, and drilling, with the tactics of moving, deceiving, and harassing. These actions are sufficient to mask detection from space assets along with detection from aircraft.[46] However, if these claims are true, this means that China has the ability to task space assets to attempt detection of troops. Also, it must have trained imagery analysts in order to examine the photographs to enable them to be able to claim that they are capable of avoiding detection.

China was expected to launch a new radar remote-sensing satellite in 2002 but it is unknown whether this would be delayed. This new satellite would have both civil and military uses, and would be comparable to Canada's Radarsat and the European Space Agency's ERS 1-2 vehicles.[47] In 1995 China's civilian space spending was approximately $1.38 billion, with around 200,000 people employed in both civil and military space programs.[48] These figures are not comparable to Western budgets and work force levels since Chinese labor costs are minimal and China does not operate a market economy.

China in 1996 began examining the feasibility of establishing a satellite navigation and position system. The Secretary General of the Science and Technology Commission, at the Chinese Academy of Space Technology declared in a statement that, "The development of an autonomous satellite navigation system is urgent. The national defense construction departments also urgently require applied satellites."[49] This system would consist of a twin-satellite navigation and positioning system consisting of two geostationary orbit satellites.[50] This proposal has sometimes been referred to as Twin Star,[51] also known as the Beidou Navigation System. Such a satellite system would have both civilian and military applications. On 31 October 2000 China launched an indigenous navigation position satellite.[52] The satellite was developed under the auspices of the Research Institute of Space Technology and was launched from the Xichang Launch Center. The second satellite of the system was launched on 21 December 2000.[53] The follow-on system is under design and consists of four satellites in geosynchronous orbit.[54] However, unlike GLONASS and GPS these geosynchronous satellites only provide regional navigation information. They are unlikely to offer precise enough information for use on missile systems.[55] The Chinese Aerospace Corporation (CASC) is examining two designs to provide a more global navigation system. One such design proposes a design of five satellites in five orbital planes at an inclination of 43.7°, whereas another envisages seven satellites in seven orbital planes at an inclination of 61.8°.[56]

The first of a new generation of communication satellites, the Dongfanghong 3 was launched in November 1994, but the satellite malfunctioned after it had attained orbit. A replacement was launched in 1997.[57] The Asiasat and APStar series of satellites are owned by Asia Satellite Telecommunications Company and Asia Pacific Telecommunications Satellite Company respectively. These two companies are partially owned by PRC interests and are based in Hong Kong.[58] China has additionally purchased communications satellites abroad to service its domestic needs. The first successful launch of such a communications satellite occurred on 30 May 1998. The satellite, the ChinaStar 1 (Zhongwei 1), was manufactured by Lockheed Martin. Prior to that on 18 August 1996 a communications satellite manufactured by Hughes Space and Communications Company failed to reach its intended orbit due to a third-stage failure. The first European built satellite, SinoSat-1 was built for the German-Chinese consortium EuraSpace with the French company Aerospatiale as the prime contractor. The SinoSat-1 is managed by the Chinese company Sino Satellite Communications Company and was launched in July 1998.[59]

In 1998 China had amassed eleven communication satellites but had allotted only limited channels to the PLA.[60] To rectify this situation a network of defense satellite communications has been proposed. The network would emphasize developing small mobile stations. This would not only expand the strategic communication channels and reception points but the mobility of the stations would allow them to be deployed to sensitive areas as part of the rapid response units to meet that requirement.[61] The Feng Huo-1 (FH-1) satellite was launched in January 2000, the first of several military communication satellites that consists of the Qu Dian command, control, communications, computer, and intelligence C^4I system.[62] This network will enable PLA commanders to communicate with their in-theater forces in near real time.

China has an interest in land remote-sensing satellites. The term remote-sensing refers to any sensing of the Earth and its atmosphere including weather satellites. An imagery satellite requires the same basic technical characteristics for civilian purposes as it does for military uses. The main difference is the resolution required, although the commercial sector is rapidly catching up to the military requirements. In 1981 with the approval of the State Council the National Remote-sensing Center (NRSC) was founded to develop and apply remote-sensing.[63] China purchased a Landsat receiving station from System and Applied Science Corporation in the United States in 1986[64] and has subsequently upgraded it several times in order to enable receipt of imagery from other satellites. This receiving station is located in Beijing. In 1989 the station covered 80 percent of China's land area with an antenna elevation above 3°.[65]

In the late 1980s China began to consider developing its own Earth observation satellites that could be used for both civilian and military meteorological purposes. As part of the Seventh Five-Year Plan the Feng Yun 1-1 (FY1-1) designed to operate in polar orbit was launched on 6 September 1988 and the Feng Yun 1-2 on 3 September 1990.[66] These satellites were set at an orbital altitude of 900km and an inclination of 99°, in order to achieve optimal observation of China's land and ocean. The remote-sensing capabilities include a multichannel scanning radiometer that has four visible bands including one infrared band, along with both analog and digital data.[67] The analog data is transmitted to Earth through VHF and the digital in S band. The data acquisition systems, in particular the 2-channel infrared scanner (DS-1230), multispectral scanner (DS-1260), metric cameras (RMK, RC-10) and synthetic aperture radar system have been imported.[68] Two satellites were imported from Spar Aerospace of Canada.[69] The satellites not only provide imagery in different spectra, but a radar satellite will enable all-weather, around the clock coverage. Synthetic aperture radars can be used for digital terrain mapping, surveillance, and target acquisition, especially over oceans.[70] However, the FY1-1 satellite operated for only thirty-nine days of its planned one year and FY1-2 lost attitude control five months into its orbital life and eventually was lost due to radiation.[71] The follow-on series of satellites, the Feng Yun 2, were designed to operate in geosynchronous orbit. This orbit could be used since the satellites using synthetic aperture radars do not need to be in low earth orbits as do photographic imaging satellites. The first was lost due to an explosion during ground processing, however, the second, FY-2B was successfully placed into orbit on 10 June 1997.[72] Another earth observation satellite, Feng Yun 1C was launched on 10 May 1999 in a sun-synchronous orbit.[73] The purpose of these satellites was announced by the Chinese for environmental disaster monitoring, however, one of the satellites was built for the Commission on Science, Technology, and Industry for National Defense.[74] This, combined with the radar remote-sensing satellites adds up to a significant reconnaissance capability.

China's International Cooperation and Its Current Space Forces

China is involved in cooperation with Brazil to build two remote-sensing satellites, the China-Brazil Earth Resources Satellite (CBERS-1 and 2). This project began on 4 September 1988 and was signed in Beijing by three representatives from the Chinese Academy of Space Technology (CAST) and three from Brazilian Space Research Institute (INPE). The ratio of responsibility of the project is split 70 percent of the costs for China, which amounts to around $150 million and 30 percent for Brazil.[75] The remote-sensing system was designed to carry three imaging systems including a charge-coupled device (CCD) that will provide 20 meters resolution which could be useful for both civilian and military purposes.[76] The CBERS project enabled China to acquire from Brazil the CCD device chamber, computers and test systems,[77] along with the wide field imager.[78] China provides the bus, panchromatic/multispectral CCD cameras, and IR sensor.[79] Along with the other reasons for cooperating with Brazil, such as reducing development time and financial costs the most important was that Brazil could act as a conduit to enable China to acquire technological know-how from other countries which China could not otherwise obtain.[80] In order to alleviate concerns that Brazil and China would cooperate in the future on military space assets, Brazil on 10 February 1994 agreed to adhere to the rules of the MTCR.[81]

The initial launch date for CBERS-1 was scheduled for December 1992, however this date was to prove elusive. This slippage was due to the fact that costs began to rise and the project did not receive the highest priority within the Brazilian government.[82] Also, the Taiyuan Satellite Launch Center does not usually function at that time of the year.[83] The CBERS-1 (known as Zi Yuan-1, ZY-1) was launched on 14 October 1999 and has a lifespan of two years, although it is extendable.[84] The data from the satellites is sent to the three ground stations in China. CBERS-2 was originally scheduled for launch in October 2001, but this date has slipped to 2002.[85] The onboard camera permits up to fifteen minutes of CCD camera data to be stored for retransmission.[86] A memorandum between Brazil and China concluded on 20 September 2000 added two further satellites to the two already contained in the agreement. These additional satellites are to have improved spatial resolution down to 5 meters, and also include greater redundancy in order to prevent electronic failure which occurred to CBERS-1 six months into its operation.[87] These satellites had an anticipated launch date of 2003, 2004, and 2006 respectively.

The HY-1 is an oceanography satellite equipped with a color ocean scanner and set to be launched in the year 2002.[88] There are two more weather (oceansats) satellites to be added in 2003-2004. The combination of the FY, HY, and CBERS satellite constellation will have an average revisit time of twelve hours.

In October 1992, China launched a FSW-1 recoverable earth imaging satellite from the Jiuquan launch site in north central China. The satellite which was placed into orbit aboard a Long March 2C returned its nose mounted reentry vehicle to earth after around two weeks. The film camera imaging system operated by the FSW-1 had both Earth resources and military reconnaissance capability.[89] The follow on series FSW-2 had a resolution of 10 meters, and could handle 2000 meters of film. The first was launched in August 1992, with others following in 1994 and 1996.[90] This series of satellite has an advanced maneuvering capability. The FSW-3 is anticipated to have a recoverable system with a one meter resolution.[91]

In 1994 China Aerospace Company and Germany's Deutsche Aerospace (now DASA) formed a joint venture, Euraspace, to build remote-sensing and communications satellites. The first communications satellite as mentioned previously, was launched in July 1998. Another joint venture was initiated in 1996 to build satellite electronics and ground facilities. This venture was between China's Xi"an Institute of Space Radio Technology and Canada's Com Dev International, and was subsequently called Com Dev Xi"an.

In May 1990, Chinese and Russian industrial representatives signed an agreement to cooperate on ten projects together. These areas of space cooperation were: satellite navigation; space surveillance; propulsion; satellite communications; joint design efforts; material; intelligence sharing; scientific personnel exchanges; and space systems testing.[92] China and Russia signed a protocol in 1994 for space cooperation. The areas reportedly of interest in cooperation were related to China's interest in human spaceflight. In April 1996 President Yeltsin signed a joint understanding on space cooperation with Chinese space officials that included training Chinese specialists at Russia's cosmonaut training facilities at Star City. By the end of 1996 the Russian Space Agency was reported to be close to signing contracts with China concerning commercial spacecraft launches, although it is unclear what was the outcome.[93] Similarly details regarding a framework agreement between China and Kazakhstan in 1998 remain elusive. The assistance from Russia and its republics provides the greatest source of space cooperation for China, more than the other countries with which China is in partnership with. This is because Russia does not place significant restrictions on its cooperation.

China's cooperation with the United States on a governmental level in the area of satellite technology has been chequered. The first delegation of space officials from China to visit NASA was in December 1978. The priority areas the Chinese identified were the purchase of a domestic communications satellite system and access to NASA's earth resource sensing satellites.[94] The Chinese announced its intent to market its Long March 2C and 3 satellite launches on the international market in 1985.[95] This forced the Chinese government to focus its attention on its relations with the United States.[96] Virtually all commercial communication satellites are manufactured by U.S. companies or include U.S. components which has meant that export licenses are required. The importance to China of the development of its satellite launcher service can be seen from the following quote:

> China is the third country in the world to offer an international satellite launching service. China's satellite industry is important for the country as it seeks to build international cooperation and new patterns of development.[97]

The first export license requests were made to the State Department in 1988 for two Australian satellites, AUSSAT-1 and 2 built by Hughes Space and Communications Company and one satellite Asiasat-1 based in Hong Kong. The Reagan administration approved the export of the three satellites in September 1988, on the proviso that China signed three international treaties: the liability for damage from space launches; negotiation of a free trade agreement with the United States concerning launch services; and an agreement protecting technology transfer while each satellite was in China.[98] The technological know-how to which the Chinese were to be confined was what they required in order to be able to launch the satellite. Included in the agreements was a quota limiting China to no more than nine international satellites between 1989 and 1994[99], and that its launch prices would be similar to other launch service providers. The approval for the export of the Australian satellites was granted by the Coordinating Committee on Multilateral Export Controls (CoCom). The Chinese government met all of the required conditions as laid out in the Memorandum of the Agreement on Satellite Technology Safeguards Between the Governments of the United States of America and the People's Republic of China which was signed in December 1988.[100]

However the events in Tiananmen Square led President Bush to suspend their export licenses. The prior conditions set out by the Reagan administration were superseded by Congress in the FY1990 Commerce, Justice, State and Judiciary Appropriations Act and the 1990-1991 Foreign Relations Authorization Act.[101] The new conditions Congress set out to the President were first, that China had achieved political and human rights reforms and second, that Sino-American cooperation was in the national interest of the United States. In December 1989, President Bush notified Congress that it was in the national interest to export the AUSSATs-1 and 2 and Asiasat-1. The first satellite that was launched was Asiasat-1 on April 7, 1990 and the AUSSATs-1 and 2 were subsequently renamed to Optus B1 and Optus B2 and launched on 13 August 1992 and 21 December 21 992 respectively.

The license agreements that China required in order to launch foreign satellites on their territory were temporarily revoked by the United States on 16 June 1991. This action was taken since the State Department had identified CGWIC as one of two organizations that had been involved in the proliferation of missile technology.[102] The relevant pieces of legislation under which the sanctions were authorized were the Arms Export Control Act and the Missile Technology Control Regime (MTCR). The invocation of the MTCR in relation to satellite launch technology rested on the interchangeability of ballistic missiles and satellite launch vehicles. China in the face of this action subsequently agreed to abide by the MTCR and the sanctions were lifted in March 1992.

A new seven-year commercial space launch agreement was signed between the United States and China in January 1995. This agreement permitted China to launch eleven foreign satellites into geostationary orbit along with the prior four launches they had previously agreed to in 1989.[103] A further five satellite launches were to be allowed,[104] on the proviso that China agreed not to price its launch services more than 15 percent below that of Western companies. If this occurred a U.S. review would be undertaken. Prior to this in the late 1980s China was charging 20 percent less than the prevailing international rates.[105] The United States strategy in the international launch service market is based on the notion that China, Russia, and Ukraine should not win a combined total of contracts of more than 50 percent of any low earth orbit satellite constellations. The 15 percent pricing policy was grounded in this strategy. A more formal pricing arrangement was reached on 27 October 1997 for Chinese space launches to low earth orbit with China agreeing to pricing its launches at the same level as their U.S and European counterparts, with adjustments made for certain differences.[106]

The procedure for issuing license agreements for the export of commercial satellites was altered in 1996 with the Commerce Department taking overall responsibility away from the State Department. This action was seen by some as a relaxation of the procedure, since the Commerce Department's overall concern is increasing trade, while the State Department's concern is not driven to the same extent in that direction. Indeed, prior to this shift the Commerce Department's licensing responsibilities included items not on the Munitions Control list such as cleaning agents, bottled gases, and spacecraft handling equipment. In 1997 the possibility arose that Loral and Hughes had transferred technology after the two companies investigated the reasons for the malfunctioning of the Intelsat 708 satellite.[107] The subsequent report by Loral which included representatives from Hughes was made available to the Chinese, before it had notified the State Department in direct contravention of the companies own policies. The companies in their defense notified the State Department of their error and claimed no technical transfer of information had been contained in the report. However, according to a leaked 1997 Defense Department report technical information that the Chinese used to improve the reliability of their warheads was contained in the company report.[108]

This evidence prompted the House of Representatives to investigate these companies and their assistance to the Chinese. This report, "The Final Report of the Select Committee on U.S. National Security and Military/Commercial Concerns with the People's Republic of China," is more commonly know as the "Cox Report." The allegations of the Cox Report remain deeply contested, although it did highlight a concern that the Chinese are using cooperation with international companies to further enhance their indigenous military space capabilities.

One author writing during the Cold War about the tensions between national security and technology transfer highlights the confusion at the center of U.S. policy:

> At times, the anti-Soviet concerns gain the upper hand, and worries about high-technology "leakage" to the Communist East lead to restrictive transfer policies to private corporations and our European allies. At other times (and even sometimes simultaneously), the national concerns for American technological and economic competitiveness make themselves felt, and policies to facilitate transfer of space technology from NASA to U.S. corporations are formulated.[109]

This quote is dated in connection with the Soviet Union, but it is however valid in connection with the recent disputes with the United States and China mentioned above. The concern with regard to China is not new. The issue is how far the shift in emphasis for technological transfer from the State Department to the Commerce Department has enabled China to use this knowledge to enhance its military space systems. For the purposes of this thesis, the Chinese intent to enhance its military space systems is the most important aspect of this episode. It can be said that the Chinese are keenly interested in using any international cooperative space programs to further their military space capabilities.

China's Interest in Antisatellite Weapons

In 1998 there was a report that China's Central Committee was giving its highest priority to the development of an antisurveillance satellite system.[110] This system is a ground-based laser generator with a capability to damage sensors of low-earth orbit imaging satellites. It is believed that the laser has been developed in cooperation with Russia, however this is not confirmed and it would be underestimating the capabilities of Chinese specialists to merely attribute its development to Russia.

The Cox Report in 1999 also addressed the issue of China's interest in antisatellite weapons. It judged that China had the technical capability to develop the CSS-2 into a direct ascent antisatellite weapon, similar to the Soviet ASAT system that used the SS-9.[111] Other reports have mentioned the possible modification of China's solid fuelled missiles, the DF-21 or DF-31 as a direct ascent kinetic kill weapon. One author has mentioned the use of the Long March 1 in an ASAT role.[112]

A website detailing the PLA's ground-based laser ASAT was posted on the Internet.[113] The information contained on the website was obtained from a report that was made available only to Congress and was not published externally. The source of the report was made available from Representative Rohrabacher. From the report the following table shows China's requirements for the system as follows:

China's Rough Estimates for Laser Weapon Systems Requirements

	Antisatellite Weapon
Operational Range (kilometers)	500-1000
Target Speed (second/second)	8
Target Hard Kill Irradiance (Watts/Centimeter)	10 x 3
Laser Brightness (Joules/Steradian)	2J x 10x17-10x19
Dwell Times (Seconds)	1-10
Average Power on Target (Watts)	10x6-10x7
Beam Quality Requirements	2
Laser Wavelength (Micrometers)	Approx. 1
Output Beam Diameter (Meters)	4
Pointing and Tracking Requirements (Microradians)	Less than 1
Adaptive Optics Requirements	Many Actuators
Beacon Requirements	Multiple Beacons

According to press reports from the Hong Kong-based Chinese newspaper Sing Tao of 5 January 2000 China has developed and ground tested an advanced antisatellite weapon, called a parasitic satellite.[114] A "parasitic satellite" is a microsatellite that is designed to attach itself to a target satellite and can be activated when required, to either jam or destroy the intended satellite. It is claimed that the parasite satellite is able to attack satellites in low, medium, or high orbit, and is so small as to not affect the target satellite's normal functions and hence go undetected. The cost of this "parasitic satellite" is one-hundredth or one-thousandth of that of an ordinary satellite.[115]

The parasitic satellite is being developed by the Small Satellite Research Institute of the Chinese Academy of Space Technology. The system has three components: the parasitic satellite, a carrier satellite and launcher, and a ground control system.[116] Owing to the mode of operation of the parasitic satellite, it must be extremely small to conceal its existence and avoid interfering with the normal operation of the host satellite. The parasitic satellite is comprised of nanometer-sized components: solar panels, batteries, computers, CCD cameras, communications and propulsion systems, auxiliary equipment, and combat systems.[117] These components use microtechnologies which enable the satellite to weigh several kilograms to several tens of kilograms, with some only several hundred grams. This low weight is essential to the functioning of the parasitic satellite.

The technology required to use microsatellites as parasitic satellites was developed in cooperation with the United Kingdom. In 1998, Hantyen Satellite Corporation and Surrey Satellite Technology Limited signed a contract to codevelop microsatellites. Within two years, the first Chinese microsatellite was launched.[118] China's Tsinghua and Harbin Universities, in a cooperative development program with the Surrey Satellite Technology Limited, launched a micro-T spacecraft in May 2000, and announced plans to develop a 22-pound nanosatellite.[119] Indeed, the connection with the PLA and Surrey Satellite Technology Limited is further elaborated with the claim that China has made great strides in the development of microsatellites capable of performing the sort of space control functions described above.[120] Other foreign participants in assisting China's miniaturization of its satellites include the former Dornier Company, now Astrium in Friedrichshafen, Germany.[121]

The PRC's Military Satellites in Orbit

Type of Satellite	Name	Date Launched	Design Lifetime
Communications	Zhongxing 22	1/5/00	Unknown
	Zhongxing 8	1/25/00	10 years
Navigation	Beidou 01A	10/30/00	2 years
	Beidou 01B	12/20/00	2 years
Reconnaissance	Feng Yun 1C, 2B	6/10/96, 5/10/99	2 years operational
	Tsinghua	6/28/00	3 years
	Ziyuan 1, 2	10/14/99, 9/1/00	3-5 years

Source: *SIPRI Yearbook 2002* (Stockholm International Peace Research Institute: Oxford University Press, 2002) and *2002 Aerospace Source Book*, Aviation Week and Space Technology, January 14, 2002.

The PLA has outlined its mission with regard to military space consisting of two categories. The first is information supporting, and the second, battlefield combating.[122] Its initial aim with relation to military space is within the information supporting. It has further defined this mission, as intelligence, navigation/positioning, and communication. One Chinese author believes that two sides in a war will concentrate on offensive and defensive operations that are conducted from space and these will become a new aspect in future wars.[123]

Conclusion

China has a moderate military space capability. The development of satellites for communications, reconnaissance in different spectral forms such as synthetic aperture radar, and electronic intelligence and navigation, offer formidable military space capabilities. It is also actively considering the development of space weapons, in the form of antisatellite weapons. The reconnaissance capabilities in particular would allow China the ability to monitor Taiwan's defenses (and others) to be used for a possible attack if it does not adhere to the principles China has laid out for Taiwan in relation to its independence status. In addition the satellite navigation could be enhanced for use in increasing the accuracy of China's ballistic missile capabilities. The ASAT capabilities, reportedly under examination are direct ascent missiles, ground-based lasers, and a parasitic satellite with an explosive charge.

In the last couple of decades in particular, China has forged a number of international agreements in the realm of space. These are often termed as civilian space ventures, however many of them have dual use capabilities as has been evidenced. To reiterate, the miniaturization satellite technology developed from the United Kingdom could be used as part of China's parasitic antisatellite capability. Similarly, the development with Brazil of its CBERS satellites assists China in developing its photoreconnaissance capabilities. This cooperation has allowed China to develop its military space capabilities considerably quicker than it would otherwise be able to do so. China now has an array of satellites which have dedicated military purposes.

China has shown an interest in developing antisatellite weapons, and hence weaponizing space. It is safe to say that the parasitic satellite ASAT is not operational, the microsatellites that are required to carry out this mission are not at the required level of technological maturity. However, it is likely that China has acquired from the former Soviet Union the technical know-how to operate the direct ascent method of satellite negation, similar to the co-orbital method of interception. Also, the laser weapon capability of blinding low-earth orbit satellites appears to be close to fruition. Certainly there are no technological barriers to China developing such a system. The PRC is certainly seeking to weaponize space. As to what extent this remains a difficult question to answer.

Notes

1. Brian Harvey, *The Chinese Space Program: From Conception to Future Capabilities* (Chichester: Praxis Publishing, 1998), 3.
2. Harvey, 3.
3. Harvey, 4.
4. Harvey, 4.
5. Harvey, 4.
6. Harvey, 4.
7. Harvey, 6.
8. Harvey, 6.
9. Harvey, 20.
10. Harvey, 28.
11. Joan Johnson-Freese, *The Chinese Space Program: A Mystery within a Maze* (Malabar, FL: Krieger Publishing, 1998), 49.
12. Brian Harvey, 38.
13. Harvey, 50.

14. Craig Covault, "Austere Chinese Space Program Keyed Toward Future Buildup," *Aviation Week and Space Technology*, July 8, 1985, 17.
15. G. Lynwood May, "New Directions for the People's Republic of China Space Program," *Signal*, December 1987, 40.
16. Radhakrishna Rao, "China's Space Plan," *Satellite Communications*, February 1987, vol. 11 no. 2, 26.
17. Wei Long, "China Builds Advanced Spacecraft Tracking and Command Network," *SpaceDaily*, May 29, 2000, http://www.spacedaily.com/news/china-00za.html.
18. You Ji, *The Armed Forces of China* (New York: I.B. Taurus, 1999), 76.
19. Ji, 72.
20. Brian Harvey, 64.
21. Ji, 71.
22. Ji, 71.
23. Mark A. Stokes, *China's Strategic Modernisation: Implications for the United States*, (Carlisle, PA: U.S. Army War College, September 1999), 41.
24. Stokes, 41.
25. Brian Harvey, 125.
26. Gordon Pike, "Chinese Launch Services: A User's Guide," *Space Policy*, May 1991, 106.
27. Pike, 107.
28. Brian Harvey, 133.
29. Harvey, 139.
30. Harvey, 139.
31. Harvey, 140.
32. Harvey, 140.
33. Yanping Chen, "China's Space Policy - A Historical Review," *Space Policy*, May 1991, 122.
34. Anne Gilks, "China's Space Policy," *Space Policy*, August 1997, 216.
35. Yanping Chen, "China's Space Commercialisation Effort: Organisation, Policy and Strategies," *Space Policy*, February 1993, 48.
36. Chen, 49.
37. Chen, 49.
38. Marcia S. Smith, *China's Space Program: A Brief Overview Including Commercial Launches of U.S.-Built Satellites*, September 3, 1998, 2.
39. Marcia S. Smith, 2.
40. G. Lynwood May, 43.
41. Marcia S. Smith, 2.
42. Marcia S. Smith, 1.
43. Pierre Langereux and Christian Lardier, "Launch Setbacks Fail to Dent China's Space Ambitions," *Interavia Business and Technology*, December 1996.
44. "Dragons in Orbit? Analysing the Chinese Approach to Space," *National Defense University*, Washington D.C. [accessed August 4, 2001] http://www.ndu.edu/inss/China_Center/paper10.htm, 5.
45. Dragons in Orbit, 6
46. Dragons in Orbit, 9.
47. Craig Covault, "China Seeks Cooperation, Airs New Space Strategy," *Aviation Week and Space Technology*, October 14, 1996, 31.
48. Covault, 32.
49. Zhu Yilin, "Fast-Track Development of Space Technology in China," *Space Policy*, May 1996, 139.

50. Zhu Yilin and Xu Fuxiang, "Status and Prospects of China's Space Program," *Space Policy*, February 1997, 73-74.
51. Brian Harvey, 153.
52. "China Launches Maiden Navigation Positioning Satellite," *SpaceDaily*, October 31, 2000, http://www.spacedaily.com/news/001031025809.g4qwq3.ac.html. See also Cheng Ho, "First Chinese Navsat In Operation," *SpaceDaily*, November 22, 2000, http://www.spacedaily.com/news/china-00zzr.html.
53. "China Launches Second Navigation Positioning Satellite," *SpaceDaily*, December 21, 2000, http://www.spacedaily.com/news/001221032826.9hwr6xru.html.
54. Mark A. Stokes, 181.
55. Stokes, 182.
56. Stokes, 182.
57. Marcia S. Smith, 8.
58. Smith, 8.
59. Smith, 8.
60. You Ji, *The Armed Forces of China*, (New York: I. B. Taurus: 1999), 77.
61. Ji, 77.
62. Colonel David J. Thompson, "The Role of China's Space Program in its National Development Strategy," *Maxwell Paper No. 24* (Maxwell Air Force Base, Alabama, August 2001), 10-11.
63. He Changchui, "The Development of Remote-sensing in China," *Space Policy*, February 1989, 65.
64. Chengchui, 70.
65. Chengchui, 70.
66. Brian Harvery, 168.
67. He Changchui, 71.
68. Chengchui, 70.
69. Wendy Frieman, "The Understated Revolution in Chinese Science and Technology," in James R. Lilley and Daria Shambaugh, eds., *China's Military Faces the Future* (Washington, D.C.: American Enterprise Institute for Public Policy Research, Washington, D.C., 1999), 255.
70. Mark Hewish, "The Sensor of Choice: Synthetic Aperture Radar," *Jane's International Defense Review*, May 1997, 34.
71. *The Final Report of the Select Committee on U.S. National Security and Military/Commercial Concerns with the People's Republic of China*, Subsection "PRC Missile and Space Forces," (Washington, D.C.: U.S. GPO, 1999), 13.
72. Cox Report, 13.
73. *Aviation Week and Space Technology*, January 17, 2000, 159.
74. Wendy Frieman, 255.
75. Jose Monserrat Filho, "Brazilian-Chinese Space Cooperation: An Analysis," *Space Policy*, May 1997, 160.
76. Marcia S. Smith, 9.
77. Jose Monserrat Filho, "Brazilian-Chinese Space Cooperation: An Analysis," *Space Policy*, May 1997, 165.
78. Michael A. Taverna, "Pacts With China, Italy Spotlight Latin American Space Ambitions," *Aviation Week and Space Technology*, October 9, 2000, 125.
79. Taverna, 125.
80. Jose Monserrat Filho, 162.
81. Jose Monserrat Filho, 168.
82. Jose Monserrat Filho, 164.

83. Jose Monserrat Filho, "Brazilian-Chinese Space Cooperation: An Analysis," *Space Policy*, May 1997, 160.
84. Michael A. Taverna, "Pacts with China, Italy Spotlight Latin American Space Ambitions," *Aviation Week and Space Technology*, October 9, 2000, 125. See also "Defense Department Details Chinese Military Space Capabilities and Plans," *SpaceDaily*, June 28, 2000, http://www.spacedaily.com/news-milspace-00a.html.
85. Michael A. Taverna, "India, China to Expand Earth-Observing Nets," *Aviation Week and Space Technology*, October 29, 2001, 88.
86. Michael A. Taverna, "Pacts with China, Italy Spotlight Latin American Space Ambitions," *Aviation Week and Space Technology*, October 9, 2000, 125.
87. Michael A. Taverna, "Pacts with China, Italy Spotlight Latin American Space Ambitions," *Aviation Week and Space Technology*, October 9, 2000, 125. See also Wei Long, "China, Brazil Continue Remote-sensing," *Space Daily*, September 28, 2000, http://www.spacedaily.com/news/china-00zzf.html.
88. Michael A. Taverna, "India, China to Expand Earth-Observing Nets," *Aviation Week and Space Technology*, October 29, 2001, 88.
89. Craig Covault, "Chinese Space Program Sets Aggressive Pace," *Aviation Week and Space Technology*, October 5, 1992, 49.
90. Mark A. Stokes, 35.
91. Stokes, 36.
92. Stokes, 184.
93. Marcia S. Smith, *China's Space Program: A Brief Overview Including Commerical Launches of U.S.-Built Satellites*, September 3, 1998, 10.
94. Stephen M. Shaffer and Lis Robock Shaffer, *The Politics of International Cooperation: A Comparison of U.S. Experience in Space and in Security*, Volume 17 Book 4, (Denver, CO: Univeristy of Denver, 1980, 26.
95. Liu Dengrui, "China's Space Industry Forging Ahead," *China Today*, September 1996, vol. 45, no. 9, 41.
96. Radhakrishna Rao, "China's Space Plan," *Satellite Communications*, February 1987, vol. 11, no. 2, 26.
97. Liu Dengrui, 42.
98. Marcia S. Smith, *China's Space Program: A Brief Overview Including Commerical Launches of U.S.-Built Satellites*, September 3, 1998, 4.
99. "Fact Sheet on the Memorandum of Agreement Between the U.S. and PRC Regarding International Trade in Commercial Launch Services," U.S. Department of State, Washington, D.C., 1989 quoted in Gordon Pike, 112.
100. Gordon Pike, 111.
101. See Congress FY1990 Commerce, Justice, State and Judiciary Appropriations Act (P.L. 101-246, Section 902) and 1990-91 Foreign Relations Authorisation Act (P.L. 101-246, Section 902).
102. Marcia S. Smith, 5.
103. Marcia S. Smith, 6.
104. The five extra international satellite launches were to be in two phases, the first would permit two satellite launches followed by a further three.
105. Radhakrishna Rao, "China's Space Plan," *Satellite Communications*, February 1987, vol. 11, no. 2, 26. The author however makes a caveat that this figure is according to Chinese official sources. China was charging $20 to $25 million to launch a communications satellite, Ariane would charge $30 million and the shuttle would cost $20 million plus $10 million for payload assistance to place the satellite into the required orbit.
106. Marcia S. Smith, 7.

107. Marcia S. Smith, 7.

108. Jeff Gerth and Raymond Bonner, "Companies Are Investigated for Aid to China on Rockets" *New York Times*, April 4, 1998 A1.

109. Walter J. Jones, "National Security, Technology Transfer Controls, and U.S. Space Policy," quoted in Daniel S. Papp and John R. McIntyre, *International Space Policy: Legal, Economic, and Strategic Options for the Twentieth Century and Beyond* (New York: Quorum Books, 1987), 66.

110. Paul Beaver, "China Develops Antisatellite Laser System," *Jane's Defense Weekly*, December 2, 1998, 18. See also Paul Richter, "China May Seek Satellite Laser, Pentagon Warns," *Los Angeles Times*, November 28, 1998.

111. "The Final Report of the Select Committee on U.S. National Security and Military/Commercial Concerns with the People's Republic of China," Subsection "PRC Missile and Space Forces,"
http://www.cnn.com/ALLPOLITICS/resources/1999/cox.report/missiles/page1.html, 10.

112. Mark A. Stokes, 186.

113. http://www.softwar.net/plasat.html.

114. Al Santoli, "PLA Successfully Tests Advanced Antisatellite Weapon," *China Reform Monitor*, American Foreign Policy Council, Washington, D.C., No. 355, January 17, 2001 http://www.afprc.org/crm/crm355.htm.

115. Santoli.

116. Cheng Ho, "China Eyes Antisatellite System," *Space Daily*, July 8, 2000, http://www.spacedaily.com/news/china-01c.html.

117. Ho.

118. Summary of the Center for Security Policy's High-level Roundtable Discussion of: 'space Power: What is at Stake, What will it Take?" 11 December 2000.
902 Hart Senate Office Building,Washington, D.C. http://www.centerforsecuritypolicy.org/papers/2001/01-P04at.shtml.

119. Al Santoli, *China Reform Monitor*, American Foreign Policy Council, Washington, D.C., No. 383, May 14, 2001 http://www.afpc.org/crm/crm383.htm.

120. Frank J. Gaffney, Jr., "Wake-Up Call on Space," CNSNews, January 9, 2001 http://usconservatives.about.com/blc0109space.htm.

121 Phillipe Cosyn, "China Plans Rapid-Response, Mobile Rocket, Nanosatellite Next Year," *SpaceDaily*, May 1, 2001, http://www.spacedaily.com/news/china-01zc.html.

122. You Ji, *The Armed Forces of China*, (New York: I.B. Taurus, 1999), 84.

123. Chen Huan, "The Third Military Revolution," in Michael Pillsbury, ed., *Chinese Views of Future Warfare*, (Washington, D.C.:, National Defense University, 1998).

Chapter Five

The United States and Soviet ASAT Programs

> While space itself is relatively remote from human conflict, certain kinds of satellite could have a potentially decisive impact on the outcome of conflicts on earth. Both sides recognize this fact. In peacetime, their satellites operate freely. But each side maintains some capability to interfere with or attack satellites that—given the outbreak of war—might threaten to reveal the location, size or readiness of their terrestrial or maritime forces.[1]

This chapter first examines the United States policy during the Cold War toward antisatellite weapons. Initially it analyzes the philosophy which believed that ASAT weapons had a destabilizing effect on the United States' relationship with the Soviet Union. The chapter then addresses the successive administrations" policies toward ASAT weapons and discusses the technological systems. The chapter also outlines the development of the Soviet ASAT in terms of both the organizational structure and eventual ASAT testing and the development of its capability. It analyzes the Soviet ASAT testing methods to gain an insight into the strengths and operational capabilities of its program. Toward the latter period of the Cold War in the late 1970s ASAT arms control measures began to be debated. These ASAT arms control measures are also analyzed. With the end of the Cold War the ASAT issue has not gone away. Indeed the issue has risen to the fore, especially in the light of U.S. policy which seeks to control space. The final section addresses the U.S. approach to this. Having analyzed the extensive Soviet ASAT development during the Cold War the chapter analyzes Russia's continuing work on ASAT weaponry.

Antisatellite Weapons and Strategic Stability

Antisatellite weapons are sometimes deemed to have a similar impact on strategic stability as ballistic missile defense, in that they are seen as destabilizing.[2] ASAT weapons threaten the satellites which are said to enhance strategic stability namely early warning satellites, communication satellites, and photoreconnaissance satellites. Early warning satellites are vital to strategic stability in that they provide warning of an impending attack, especially in a nuclear context. In a nuclear arena warning time is essential to strategic stability in that it prevents one side achieving a surprise first-strike attack on the other side's more vulnerable retaliatory nuclear assets, namely ICBMs and nuclear-equipped aircraft. An ASAT capability targeting early warning satellites is seen as extremely destabilizing in that it undermines a central essence of nuclear deterrence, namely that a surprise attack is unachievable. Also, the targeting of photoreconnaissance satellites which are important in the context of arms control verification, undermines the stability of the international security environment which arms control can provide. It is for these reasons that ASAT weapons are deemed to be destabilizing in an international security context.

The United States ASAT Program during the Cold War

The Eisenhower administration's position toward the development of an antisatellite system was founded on the belief that the United States was more reliant on reconnaissance information provided by satellites than the Soviet Union and subsequently did not want to initiate anything which could jeopardize that. This was because of the closed nature of the Soviet society that did not allow the United States to gain information concerning it, whereas the U.S. society was a very open and provided a great deal of information which the Soviets were able to use. Indeed, the following quote from Herbert York, the former Director of Defense Research and Engineering, encapsulates this belief:

> The President himself, in recognition of the fact that we didn"t want anybody else interfering with our satellites, limited [one ASAT] program to study only status and ordered that no publicity be given either the idea or the study of it.[3]

Implicit within this belief was that the United States, by forgoing the development of an ASAT capability, would have a subsequent effect on the Soviet Union's own desire to have such a system. Indeed this policy flowed from the sanctuary school of space policy, despite the fact that the Eisenhower administration was seen to be "hedging its bets" by pursuing the conceptual development of an ASAT system.

The Kennedy administration's like Eisenhower's, was willing to authorize the development of other ASAT programs in the eventuality of an unforeseen Soviet space threat. In February 1963 the Kennedy administration published the following statement by Marshal Biriuzov, chief of the Soviet Rocket Forces: "It has now become possible to command from earth to launch missiles from satellites at any desired time and at any point in the satellite trajectory."[4] Indeed further evidence of Soviet intentions was provided by Secretary of Defense McNamara testifying before Congress a month prior: "the Soviet Union may now have or soon achieve the capability to place in orbit bomb-carrying satellites . . . [and] we must make the necessary preparations now to counter it if it does develop."[5] This had led McNamara to instruct the U.S. Army in May 1962 to develop and modify NIKE-ZEUS in an ASAT role.

The ASAT was Program 505 which was an adaptation of the Army's NIKE-ZEUS Anti-Ballistic Missile; work began in 1955. The technology of an ABM is able to be adapted for an ASAT role since both missiles are intended to intercept targets in space. The differences however lie in the geometry of the interceptions; an ICBM observes a curved trajectory compared to the horizontal path of a satellite. There are also differences in angle, distance, and speed of the target which must be adjusted by the guidance radar. An ABM also has to be operational at all times. In many ways an ASAT capability is less demanding than that of an ABM since only one target would be engaged at a time. The target's flight path, direction and altitude would be known well in advance, whereas an ABM has to contend with multiple targets, decoys, jammers, and booster debris simultaneously, with little warning time.

The first NIKE-ZEUS ASAT test successfully intercepted an imaginary target in space at an altitude of 100 nautical miles (185km), and was within the lethal distance of the nuclear warhead.[6] A second ASAT test was conducted and intercepted an imaginary target in space at a range of 151 nautical miles (279km). However several failed tests followed and Program 505 was eventually phased out in 1966, while the rival Air Force Program 437 received the ASAT mission.[7]

Program 437 used a Thor intermediate-range ballistic missile to reach its target. The first intercept by Program 437 occurred in February 1964. The Thor missile was launched at a target, a Transit 2A rocket body, occupying a 564 by 335 nautical mile orbit, inclined at 66.7 degrees. The intercept point was at an altitude of 540 nautical miles, and the warhead passed close enough to the target to be considered a successful interception.[8] The following two tests were also deemed to be successful. In June 1964 the Thor ASAT system was declared operational. However, the effects of using a nuclear armed ASAT was considered in terms of its nondiscriminating effect on friendly satellites and the search for a nonnuclear ASAT began. Consequently on 1 April 1975 Program 437 was terminated partly due to its inability to deal with the threat from the increasing number of Soviet military satellites.[9]

The United States position for a U.S. ASAT capability in the 1970s was that the United States should match the Soviet ASAT capability as a means of deterring attacks on U.S. satellites. A weakness of this position was that it was argued that the United States was more dependent upon its satellites for military effectiveness than the Soviet Union, therefore in a tit-for-tat exchange the United States would be in a weaker position. The United States would have been in a weaker position vis-a-vis the Soviet Union since the Soviets had fewer military forces deployed beyond their borders and could rely on ground-based lines of communications, as well as the fact that the Soviets had less need for worldwide communications and navigation aids.[10] For these reasons it is unlikely that the possession by the United States of an ASAT capability would have acted as a deterrent for the Soviet Union making use of its ASAT capability.

The outgoing Ford administration and the incoming Carter administration recognized the Soviet "antispace defense" system as a threat to U.S. space assets. In response to this newly perceived threat the Ford administration planned to "increase significantly the U.S. space defense effort over a broad range of space-related activities which include space surveillance, satellite systems survivability, and the related space operations control function (meaning a U.S. ASAT)."[11] The Carter administration, unlike any of the previous administrations was faced with the likelihood of a Soviet ASAT becoming operational. However, like the Eisenhower and Kennedy administrations, the Carter administration pursued a policy of negotiating arms control on ASATs while maintaining research and development of an ASAT system as insurance. Research and development into ASAT technologies was heavily constrained. The following quote from the Carter administration's Secretary of Defense Brown encapsulates the essence of this policy:

> As the President has clearly stated, it would be preferable for both sides to join in on an effective, and adequately verifiable ban on antisatellite (ASAT) systems; we certainly have no desire to engage them in a space weapons race. However, the Soviets with their present capability are

leaving us with little choice. Because of our growing dependence on space systems we can hardly permit them to have a dominant position in the ASAT realm. We hope that negotiations on ASAT limitations lead to a strong symmetric control. But in the meantime we must proceed with ASAT programs (for the present short of operational or space testing), especially since we do not know if the Soviets will accept the controls on these weapons that we would think necessary.[12]

The Carter administration engaged the Soviets on three separate occasions from 1978 to 1979 in the pursuit of an ASAT limitations treaty, but the Soviets were unwilling to come to an agreement during this era.[13] The Soviet invasion of Afghanistan put an end to the negotiations in late 1979.

There were several key events early in the Reagan administration that polarized the ASAT issue. The Soviets in 1981 submitted to the United Nations a draft ASAT treaty calling for the banning of weapons in space. Soviet Premier Andropov two years later continued the Soviet "peace initiative" by denying Soviet first use of ASATs in outer space and offered to dismantle the existing Soviet ASAT system and prohibit further development. In 1983 the Soviets proposed another draft treaty to the UN. This called for "a ban on the use of force in space and dismantlement of existing ASAT systems."[14] The Soviets at the same time as these proposals tested their co-orbital ASAT system for its twentieth and final test in 1982. The Soviet Union shortly after unilaterally declared a moratorium on any further tests of its co-orbital ASAT.[15] This unilateral declaration of a moratorium provided ammunition for opponents in Congress of the U.S. development and deployment of an ASAT. However, the Reagan administration determined that the development, procurement, and deployment of a U.S. ASAT was vital to national security interests despite the Soviet proposals.

The Reagan administration showed strong support for a U.S. ASAT capability by requesting additional funding from Congress from fiscal year 1982 through to fiscal year 1985. Indeed, Congress appropriated each year what the administration requested.[16] In addition to the funding increases the Reagan administration provided a rationale for the acquisition of an ASAT capability. In a report to Congress on 31 March 31 1984, President Reagan cited two primary reasons for pursuing a U.S. ASAT. First, a U.S. ASAT capability to destroy satellites was required to deter Soviet attacks on U.S. satellites in a crisis or conflict. The policy statement cited the example that if the Soviet Union used its ASAT capability in a crisis or conflict to disable or destroy a U.S. satellite, the United States would have no means to respond in kind to avoid escalating the conflict.[17] Second, it was argued, "a comprehensive ASAT ban would afford a sanctuary to existing Soviet satellites designed to target

U.S. naval and land conventional forces."[18] Therefore, the Reagan administration argued a capability was required "for U.S. and Allied security to protect against threatening satellites."[19] The Reagan administration's policy for an ASAT requirement was thus as a means to deter the Soviets from using their co-orbital ASAT to attack U.S. space systems and a means to negate Soviet space systems designed to target U.S. forces. The Reagan administration's policy toward ASAT represented a departure from the policies espoused by previous administrations in that arms control measures were no longer deemed desirable. This was mainly due to the ideological standpoint of the administration which did not see space as being different to any other geographical environment and hence free from weaponization. The twin-track policy under the Eisenhower and Carter administrations of simultaneously pursuing research and development and ASAT arms control measures was effectively over.

The air-launched U.S. ASAT capability began in the early 1970s with full-scale development commencing in 1977. The system involved the "direct ascent" of an interceptor to its target, in contrast to the Soviet co-orbital ASAT system tested between 1968 and 1982. The interceptor consisted of a miniature homing vehicle (MHV) on a two-stage missile. The system was mounted under an F-15 fighter, the SRAM first stage and the Altair second stage would have taken the interceptor up another 500 kilometers, from where the MHV would use its eight heat-seeking infrared sensors to acquire the target, and then fire small rocket thrusters to ram the target. Destruction was to be achieved by velocity impact.

The weapon was launched from information supplied by the ground-based satellite tracking network.[20] Homing was to be achieved through a combination of eight infrared telescopes, a set of small thrusters and a laser gyroscope. The infrared sensors identified the target against the cold background of space. This guaranteed accurate data and prevented the miniature vehicle from "attacking" stars.[21] It spun at twenty revolutions per second, which not only kept it stable but assisted the eight telescopes in acquiring and locking on to the target. The cylinder rotated while the gyroscope determined when the various thrusters were to be fired in order to bring it into the path of the target. The outer shell of the miniature vehicle was composed of fifty-six small cylinders of solid rocket-propellant, the nozzles of which pointed out to the side. When fired, under control of the guidance system, they moved the vehicle body to keep it on a collision course. The rockets were fast-burning so as not to upset the spin stabilization. The guidance task of firing the correct rocket at the proper time was a major one requiring extremely sophisticated electronics, and timing was of the essence because of the vehicle's fast spin rate.[22] After the miniature vehicle's course had been corrected, counterfiring stopped the lateral drift. To achieve accuracy a laser-gyro acted as a clock enabling the onboard computer to determine which rockets have fired—they were single-shot only—and allowed the miniature vehicle to rotate past the spent rockets. Additional rockets were used to prevent the miniature vehicle developing "wobble" due to the firings. The ensuing high speed collision destroyed the target. The energy of such an impact is akin to hitting a satellite with a shell from a battleship's main gun.[23] A direct collision at such high velocity was simpler than fusing and exploding a warhead.

In principle any F-15 could have been adapted to carry the antisatellite weapon. Carrier-based F-14s or midair refueling of the F-15 would have enabled the antisatellite weapon to attack almost any position in the world.[24] This would have allowed the U.S. to target all of the low-orbit satellites along with the highly elliptical orbit satellites known as Molniya satellites, possessed by the Soviet Union.

The first flight test occurred on 21 January 1984, when an ASAT booster, without a MHV aboard, was launched against a point in space. A second test in November was targeted against an infrared emitting body (a star) to test the ability of the MHV to distinguish between its target and the background infrared emission of space. Both of these were considered to be successful. The third, and most important, test took place on 13 September 1985; in this test, the complete system was launched against a target satellite. The MHV successfully intercepted the target.[25] The miniature vehicle was destroyed by direct collision with the target at 45,000 feet per second (13,716 mps).

The F-15-launched antisatellite missile was a two-stage solid-fuel rocket 17.75 ft (5.4m) long and weighed 2632 lb (1194 kg). The first stage, 17.6 in (6.9 cm) in diameter, was based on the Boeing Short Range Attack Missile (SRAM). At the base are two small fixed fins and three large movable fins which control the vehicle during atmospheric ascent.[26] The second stage had an Altair III rocket motor of the kind used as the fourth stage on the Scout launch vehicle. It was specially strengthened for its antisatellite role and was fitted with small hydrazine thrusters for attitude control. The second stage was 19.76 in (7.8cm) in diameter. At its forward end was the miniature vehicle, with its spin table and subsystems (such as the inertial reference unit, computer and cryogenic tank for cooling the infrared sensor). An inertial guidance unit provided control during powered flight until a specific point in space was reached.[27] At this point, the miniature vehicle begins to search for its target. After second-stage burnout, it spins up and the target satellite was acquired.

The F-15 launch aircraft itself required certain modifications. An electronic package replaced the 20mm ammunition container. There were wiring charges and a special centerline pylon which included a microprocessor, a communications line between the missile and aircraft, a back-up battery, electrical connections, and a gas generator ejection system. The pilot's launch duties were minimal as he receives steering commands via the cockpit head-up display. For most attack profiles, the ASAT was launched while the F-15 was in subsonic, straight, and level flight. For satellites in higher orbits, a supersonic climb would be used. This added speed to the ASAT and avoids the need for a sharp pullup which might overstress the missile. The launch was automatic with a ten to fifteen second window.[28]

The F-15 ASAT has a number of advantages over a more conventional system. An F-15 could be flown to wherever necessary to accomplish an interception. A fixed-based ASAT, dependent on a large rocket, lacked such flexibility. As long as there are F-15s, ASAT missiles, supplies, and means to function in comparison with a fixed-base ASAT, the air-launched ASAT would be a candidate for attack during the early stages of an escalating war.[29] It was economically feasible to build enough of the weapons to cope with a high enemy launch rate.

In the mid-1980s Congress began constraining the U.S. ASAT program. This was due to the difference in ideology with regard to the weaponization of space between the Democrat majority and Republican minority. The fiscal year 1986 appropriation procurement money was significantly slashed and in fiscal years 1987 and 1988 Congress denied procurement funds completely. On 19 December 1985 a Congressional ban prohibited any further tests of U.S. ASATs in space until and unless the Soviets tested its ASAT again. In fiscal years 1987 and 1988 Congress continued the ban.[30] Indeed, an Office of Technology Assessment report highlighted the complexity of the ASAT issue:

> In choosing between ASAT weapon development and arms control, one wishes to pursue that course which makes the greater contribution to U.S. national security. This is often characterized as a choice between developing a capability to destroy Soviet satellites while assuming U.S. satellites will also be at risk, or protecting U.S. satellites to some extent through arms control while forfeiting effective ASAT weapons. The better choice could, in principle, be identified by comparing the utility which the United States expects to derive from its military satellites with the disutility which the United States would expect to suffer from Soviet MILSATs during a conflict. Such a compariso although—possible in principle—is made exceedingly difficult by the number of conflict scenarios which must be considered and by the lack of consensus or official declaration about the relative likelihood and undesirability of each scenario.[31]

During the latter part of the Reagan administration Congress was unable to be convinced of the deterrent value of an ASAT. The influence on Congress of the earlier Soviet initiative for banning weapons in space and the moratorium on testing of their own ASAT system cannot be discounted. In February 1988, Secretary of Defense Carlucci announced the cancellation of the Air Force's F-15 ASAT program citing the negative impact of the Congressionally mandated ASAT test ban.[32]

The Soviet ASAT Program during the Cold War

The first indication that the Soviet Union was seriously developing an ASAT capability came in 1963-1964 with the formation under the PVO-Strany air defense branch of a special antispace defense detachment called PKO (Protivo Kosmicheskaya Oborona).[33] The mission of this unit was to repel any attack emanating from outer space.

The Soviet Union continued to proceed with an ASAT program despite the cancellation of the United States Army ASAT program in 1966 and the Air Force ASAT program in 1970. The motivation for the Soviet antisatellite program can be seen from its doctrinal concept of "antispace defense." The Soviet Union in 1965 defined this concept in the following manner:

> The main purpose of anti-space defense is to destroy space systems used by the enemy for military purposes, in their orbits. The principal means of anti-space defense are special spacecraft and vehicles (for example, satellite interceptors), which may be controlled either from the ground or by special crews.[34]

The Soviet view of the requirement for a satellite negation capability was similar to that of Generals White and Gavin for a U.S. ASAT program.

The Soviet ASAT weapons used a "hot-metal kill" weapons which was essentially an explosion in the vicinity of the target satellite which produced a spherical cloud of shredded metal expanding evenly in all directions.[35] The use of a high-explosive warhead, as opposed to a nuclear circumvented the Outer Space Treaty. However, the use of conventional means meant the ASAT had a narrow miss distance and had to pass closer to the target satellite for a successful kill.

From October 1968 the Soviet Union had tested twenty satellite intercepts against Russian target spacecraft. In sixteen of these tests the intercept distance was deemed close enough for the mission to be termed a success.[36] The interceptor vehicle was able to close in on the target satellite within one or two orbits which demonstrated a quick reaction capability. The SS-9 boosters which launched the antisatellite payloads were able to be wheeled from their shelters at the Tyuratan site and prepared for launch in less than ninety minutes.[37]

The intention of the series of Soviet satellite testing starting in 1968 only became known after six months had elapsed and the first full interceptor test had occurred.[38] On 19 October 1968 Cosmos 248 was put into orbit and on the following day Cosmos 249 was placed into an orbit that equated with the orbital plane and apogee of Cosmos 248. Indeed, within four hours a close high-speed "flyby" took place. What was more significant was that Cosmos 249 was destroyed after the flypast. Though this was not the first occasion that Soviet satellites had been exploded in orbit its occurrence with the interception of another satellite was enough to suggest the initiation of a new type of activity.[39] This was confirmed with the launch of Cosmos 252 on 1 November 1968 when it performed almost identical maneuvers to Cosmos 249, and exploded after passing close to Cosmos 248.

The reason for the destruction of the interceptor was not clear. There were however two possible theories:

> The explosions of these two payloads could mean that they carried instrumentation and other devices the Russians did not want to leave in orbit for some future generation of curious inspectors of another nationality to find; or they could have been exercising the destruct mechanism, presumably at a safe distance so as not to destroy their own target.[40]

Not until the 1983 edition of Soviet Military Power was official light shed on the matter. According to the report the Soviet ASAT detonates a pellet warhead near the target to effect a kill by damaging vital satellite components.[41]

A two-year cessation followed until on 20 October 21970 Cosmos 373 was launched from Tyuratam into an unusual orbit with a high apogee. This orbit was subsequently modified into a circular orbit similar to the first target satellite Cosmos 248. Three days later Cosmos 374 was launched from Tyuratam; this interceptor satellite was maneuvered to match the orbital altitude of Cosmos 373 at its perigee and a high-speed fly pass occurred.[42] However, Cosmos 374 apparently conducted an unsuccessful two-revolution interception with Cosmos 373.[43] The interceptor was then detonated. A week later the exercise was repeated using the same target satellite, but a new interceptor Cosmos 375 was used.[44]

In 1971 the satellite interception tests began to differ from the previous tests in a number of ways. The target satellites were launched from Plesetsk in the northwest of the Soviet Union instead of from Tyuratam, and were at a new inclination of 65.8 degrees. The launch vehicle, a modified SS-5 intermediate-range ballistic missile indicated that the target was significantly smaller than the earlier ones.[45] The new series of tests began with Cosmos 394 the target satellite, launched on 9 February 1971 and the interceptor satellite Cosmos 397 launched from Tyuratam sixteen days later.[46] The interceptor, initially in low orbit, maneuvered to a higher altitude to undertake a similar perigee-matching exercise with Cosmos 394. The interceptor satellite was detonated after the intercept. Unlike the previous intercepts, the intercept satellite was not used again for a second interception as was the case for all subsequent target satellites which were deemed successful.

On 19 March 1971 a new target satellite Cosmos 400 was launched into a circular orbit, approximately 1,000km, using a SS-5 booster from Plesetsk. The interceptor satellite, Cosmos 404, was launched from a SS-9 from Tyuratam sixteen days later. This maneuvered into a circular orbit similar to Cosmos 400. At the start of Cosmos 404's second revolution it was less than three minutes ahead of its target, and by the end of the third it was only one minute behind. With their orbital elements and hence their orbital velocities similar a much slower flyby was achieved which suggested that this mission was to test inspection equipment.[47] Instead of being detonated Cosmos 404 was deorbited back to earth. Whereas the previous five ASAT tests had ended with the interceptor maneuvering to a higher orbit and then exploding, Cosmos 404 performed a braking maneuver and reentered the earth's atmosphere.

On 29 November 1971 Cosmos 459 was launched from Plesetsk into the lowest ever orbit by a SS-5 launcher at approximately 250km at a 64.8 degree inclination. On 3 December 1971 Cosmos 462 was launched from Tyuratam into an unusual orbit and completed the familiar high-speed interception at their respective perigees.[48] The interceptor satellite Cosmos 462 was detonated after the flyby. This interception was more demanding owing to the lower altitude, which due to the higher drag of the earth makes the prediction of the speed and the likely position of the target satellite more complex. This was the last test until they resumed in 1976.

This first phase of testing demonstrated that the Soviet Union had a rudimentary yet significant antisatellite capability that threatened an important category of U.S. satellites. As one analyst observed:

> Within a period of eleven months the Russians had demonstrated their ability to place hunter spacecraft in the vicinity of targets with orbits

characteristic of [U.S.] electronic ferrets, meteorological and navigation satellites and photo-reconnaissance payloads.[49]

The first four tests of the Soviet antisatellite system had produced a 50 percent success rate which was not bad for the initial test phase of a major new weapons program. The three successful tests in the following year raised the success rate to over 70 percent and probably signaled a Soviet ASAT initial operational capability. The 1971 tests more importantly demonstrated new characteristics, particularly a flexibility in attack geometry.[50]

The Soviet antisatellite system compared with the U.S. Thor-based system had several superior capabilities. The system had considerable flexibility in its intercept trajectory, allowing attack from a number of directions and hence making countermeasures more difficult.[51] The nonnuclear warhead eliminated any possible collateral damage from a nuclear detonation which was used in the U.S. system. The reach of the Soviet system had demonstrated that U.S. military satellites in orbits below 1,000 km were vulnerable. The Soviet ASAT used the SS-9 as a launcher which was probably chosen because there was an abundant supply since it had become out of service. This was deployed in large numbers in the southern part of the Soviet Union which meant that many U.S. satellites could have potentially been negated in a short space of time.[52] Although refitting the payloads would not have been that easy as the ASAT SS-9 needed to have special facilities.

The Soviet ASAT system did possess some significant limitations. During the tests the interceptors had been placed in orbits coplanar with their targets which meant that an ASAT could be launched at a specific target only twice each day from a given launch site.[53] The flight time of the interceptors was in excess of three hours which could allow the target satellite an amount of time to deploy countermeasures. A further constraint of the coplanar attack was the inability to intercept satellites in inclinations below forty-five degrees, which ruled out attacks on U.S. manned spacecraft and other NASA satellites.[54] Finally, most U.S. satellites were in orbits above 1,000km including the early warning and communication satellites.

The Soviet antisatellite tests entered a self-imposed moratorium coincident with the birth of detente and the signing of the SALT I accords. Although at the end of September 1972 Cosmos 521 was launched from Plesetsk into an orbit characteristic of a target satellite for a future ASAT test, it was never intercepted.

The Resumption of Soviet ASAT Testing

The Soviet decision to resume ASAT testing in 1976 was multipurpose. The show of resolve displayed by the testing might have brought the United States back to the bargaining table with new concessions, as the SALT II negotiations were underway, especially since the United States' ASAT system had been dismantled. Also, if the SALT II process was abandoned with a consequent rise in international tension a Soviet operational antisatellite system might be required. The four-year cessation in testing had also allowed Soviet engineers time to design new ASAT hardware and operational options.[55] These would have to be tested before being adopted into a fully established system.

On 12 February 1976 a target satellite Cosmos 803 was launched from Plesetsk on a SS-5 into an orbit at an inclination of 66 degrees. Four days later Cosmos 804, the interceptor satellite, was launched from Tyuratam into a more eccentric orbit at an inclination of 65.1 degrees. After several maneuvers its orbit was changed and its inclination altered to that of Cosmos 803, with the interception taking place over the Soviet Union.[56] However, this test was deemed to have been a failure since the miss distance between the satellites was around eighty nautical miles and the interceptor satellite was unusually brought back to earth. Cosmos 803 was again used as a target satellite on 13 April 1976 when Cosmos 814 was launched from Tyuratam within four minutes of Cosmos 803 passing over the launch site. Tracking data showed the Cosmos 814 interceptor in an initial 297 by 72-mile orbit, which was lower than the target's 385 by 340-mile orbit. This lower orbit meant that Cosmos 814 gained on its target. Once it had caught up in this manner, Cosmos 814 fired its onboard engine and assuming an elliptical orbit made a fast flyby.[57] From launch to interception had taken forty-two minutes. The appearance of the "Pop Up" profile required less than one orbit from launch to interception and provided a fast reaction capability. Between 1976-1977 the new technique was tested in a variety of circumstances. It was tested against a target in a medium-altitude orbit, a highly elliptical orbit, and a low elliptical orbit. Each test imparted different demands on the interceptor.[58] The Soviets had demonstrated a significant new enhancement of the system with the time from launch to intercept was cut in half. Thus an intended target would receive less warning time of an attack and might not be able to employ countermeasures.

On 8 July 1976 Cosmos 839 was launched but was placed in a much higher orbit than previous target satellites. The lowest point of the orbit was nearly 1,000km above the Earth's surface, while the apogee reached an altitude of 2,100km. An interceptor satellite Cosmos 843 was launched on 21 July but it was deemed to have failed to have reached the required height and reentered the atmosphere afterwards. However, it was possible that it may not have been a failure and that the interceptor could have maneuvered close to the target shortly after launch and been recovered in less that one revolution.[59]

The Soviet Union reverted to the rapid flypast interception followed by the destruction of the interceptor satellite. This took place on 9 December 1976 with the launch of Cosmos 886, the target satellite and Cosmos 886 the interceptor satellite launched on 27 December of that year.[60] There had been four attempted intercepts in 1976, although some of them may have been failures it was the highest number of tests in any one year. A further four tests beginning in May were conducted in 1977. Further information regarding the Soviet satellite interceptor came to light. Its dimensions were between 15 to 20 feet in length, 5 feet in diameter, and weighed around 2.5 tons. It had two main boosters for maneuvering in space. It was noted that the Soviets had been experimenting with a new guidance system. Whereas previous interceptors used a radar homing system, a new optical infrared sensor was used for the Cosmos 880/866 intercept on 27 December 1976 possibly in anticipation of U.S. countermeasures.[61]

The antisatellite interceptors prior to Cosmos 886 used a radar seeker to acquire and to track the target satellite as they moved in to simulate the intercept. Cosmos 886 used a new sensor which relied on reflected sunlight or possibly the infrared emissions of the target satellite to serve as the homing device.[62] There were two principal advantages of an acquisition and tracking sensor of this type. Optical-thermal sensing systems are typically much lighter and compact and require less electrical power than radar. More importantly, optical-thermal sensors are harder to counteract. Radar seekers can be jammed by a range of electronic techniques whereas decoys are often required to fool sensors which operate in the visible or near-visible portion of the spectrum.[63]

The April 1976 test was reported by *Aviation Week and Space Technology*. This article gave readiness details of the system. It reported that between 1972 and 1975 observations had been made of ground exercises which included SS-9s with antisatellite payloads.[64]

On 19 May 1977 Cosmos 909 was launched from Plesetsk into a highly elliptical orbit at an inclination of 66 degrees. Four days later Cosmos 910 the interceptor satellite was launched from Tyuratam into an orbit with the same inclination. Instead of a fast flyby interception occurring the interceptor satellite returned to earth within one revolution. This was initially interpreted as a failure. However, when on 17 June 1977 another interceptor, Cosmos 918, was launched against the previous target, Cosmos 909, a new method of interception was apparent.

Cosmos 918 was initially launched into 197 by 124 km orbit at the same inclination as the target satellite, but in a rapid maneuver, the interceptor suddenly "popped up" to pass the target satellite at its apogee.[65] In the same movement, the interceptor returned to earth. This demonstrated a greater degree of flexibility in the use of the Soviet antisatellite system and a capability to intercept satellites at higher altitudes.[66] This method was again used against the target satellite Cosmos 959 on 21 October 1977. Cosmos 961 was launched five days later and within three hours performed a low-orbit demonstration test of the high orbit pop-up technique at an altitude of 150km.[67]

The back-to-back tests of Cosmos 918 and 961 further expanded the capabilities of the Soviet ASAT. The new orbital intercept extremes of a maximum of 1,575km and a minimum of 150km easily covered U.S. low-altitude satellites. In addition, one test demonstrated a two-revolution intercept while another fulfilled its intercept in just one revolution. This flexibility would have made U.S. decisions of what countermeasures to employ and when to activate more difficult. Countermeasures employed to combat a two-revolution intercept would be inadequate if the ASAT arrives after one orbit. Equally if the activation of the evasive maneuvers or decoys against an anticipated one-revolution intercept, then the intercept occurs an hour and a half later. The decoys may have either dispersed beyond effective limits or may have exhausted their energy sources.[68]

The Soviets then reverted back to the earlier intercept method followed by the detonation of the interceptor vehicle. On December 13, 1977 Cosmos 967 was launched from Plesetsk with the interceptor Cosmos 970 following on 21 December. On this occasion a slow flypast was completed after the original orbit of the interceptor had become circular. The interceptor vehicle was destroyed afterward. On 19 May 1978 Cosmos 1009 was launched from Tyuratam and maneuvered for a close inspection of Cosmos 967 before returning to earth. This was the last test in the series before the first round of antisatellite arms control negotiations in Helsinki.[69] Testing resumed after it became apparent that the limitation talks would not continue after the Soviet Union had invaded Afghanistan in 1980.

Since 1968 there had been three distinct types of satellite interception: the fast flypast, the slow flypast, and the "pop-up" technique. While U.S. satellites in low earth orbit were considered vulnerable, the Soviet system had not demonstrated an ability to attack geostationary orbits which contained the early warning satellites and communication satellites.

The Soviet ASAT program moved into an engineering and development testing proceeding at the rate of one flight per year during 1978-

1982. The original radar-guided ASAT interceptor was capable of intercepts to about 5,000km using a one- or two-revolution trajectory.[70] The primary Soviet difficulty focused on the development of the optical-thermal guided weapon. This device was tested between 1978 and 1982 against targets in roughly circular orbits of 1,000km altitude, and all four appear to have been unsuccessful. On 18 April 1980 Cosmos 1174 intercepted Cosmos 1171 which had been launched on 3 April. This test was deemed a failure since the interceptor did not pass closer than the 8km which was considered to be the lethal radius of its shrapnel warhead.[71] The test in 1981, Cosmos 1243, against the target satellite Cosmos 1241 indicated that there had been a possible close encounter but there was reason to believe that the intercept was not completely successful. A month later a new interceptor Cosmos 1258, believed to have been radar-guided, made an intercept against the same target, Cosmos 1241. No target had been engaged twice when the original attempt had been successful. The reversion back to a radar-guided ASAT probably reflected the consternation after the four failures with the optical-thermal system.[72] The Cosmos 1258 intercept was deemed a success.

The June 1982 ASAT test was the twentieth Soviet orbital testing in fourteen years. The target satellite Cosmos 1375 was placed in a circular orbit of 1,000 km the same as the previous five tests. On 18 June Cosmos 1379 the intercept satellite was launched on a two-revolution intercept. Despite the initial accurate orbital maneuvers the fuse failed to fire on time and the intercept failed.[73] The significance of this test was that Cosmos 1379 was part of a simulated exercise during which front-line strategic and tactical weapon systems were tested. It included ICBM, SLBM, and IRBM firings, and also ABM engagements against dedicated targets. Also, the command, control, and communications (C3) networks were tested in a simulated wartime environment along with the support radars (Hen House, Dog House, Cat House, and Try Add). In the space sector of this exercise the launch of two Soviet satellites, one photoreconnaissance from Tyuratam and one navigation from Plesetsk, was made between the launch of Cosmos 1379 and its attempted intercept of Cosmos 1375.[74] Prior to this no space launch had occurred during an ASAT test or from Tyuratam. The space launches may have imitated the orbiting of replacements for those destroyed by the United States during the simulated conflict.[75]

The Soviet ASAT system was operational in 1971. This was not confirmed until 1984, as prior to this it was believed to have become operational in 1977.[76] There were attempts to link operational intentions to the fact that since 1971 Soviet ASAT tests occurred at an inclination of 65.8 degrees and hence were aimed towards Chinese satellites. The case was made since Chinese satellites are flown nearer this inclination than U.S. satellites, and the first Chinese satellite appeared in 1970. Indeed, one commentator has argued that the Soviet interceptor tests followed both of the first two Chinese launches of satellites in April 1970 and March 1971. In addition the inclination of the interceptor was similar to that of Chinese satellites, suggesting that the intended target was the Chinese satellites.[77] However, the inclination of 68.5 degrees was a consequence of launching the target from Plesetsk (Tyuratam no longer had the facility to launch the SS-5 derived space vehicle), the launching of the interceptor from Tyuratam (until 1977 Plesetsk did not have a facility to launch the SS-9 derived space vehicle), and certain range and safety intercept restrictions.[78] The Soviet ASAT performed coplanar intercepts and subsequently the targets and ASATs had to orbit the Earth at the same inclination. The highest inclination launch from Tyuratam was 73.4 degrees and the lowest inclination launch possible from Plesetsk was 62.8 degrees, so the ASAT inclination had to fall within this.[79] The only common inclinations flown from each site have been 62.8 degrees, 65 degrees and 65.8 degrees. However, no characteristic of the Soviet ASAT was dependent upon orbital inclination, that is the launch azimuth, since this factor does not affect the intercept geometry. Therefore, the linkage between the Soviet ASAT and the Chinese 'space threat" was misleading.[80] In addition, the Soviet ASAT program was developed and tested before the first Chinese satellite flew.

The Search for ASAT Arms Control

The first notification that the Carter administration was seriously considering ASAT arms control and had proposed the issue with the Soviet Union came at a press briefing by President Carter on 9 March 1977 where he announced:

> I have proposed both directly and indirectly to the Soviet Union, publicly and privately, that we try to identify those items on which there is relatively close agreement—not completely yet, because details are very difficult on occasion. But I have for instance, suggested we forego the opportunity to arm satellite bodies, and also to forego the opportunity to destroy observation satellites.[81]

The issue was raised again by Secretary of State Vance on his visit to Moscow in March of that year. Although not the primary focus of the visit, both the Arms Control Disarmament Agency and State Department had prepared briefing papers that outlined a range of antisatellite arms control options. At the press conference following the meeting on 30 March Secretary of State Vance announced that both sides had agreed to set up working groups to discuss specific areas of arms limitation, including one for antisatellite weapons.[82] Prior to the meeting in Moscow, President Carter had issued Presidential Review Memorandum PRM/NSC-23 that directed the recently created NSC Policy Review Committee to review existing policy and formulate overall principles to guide U.S. space activities.[83]

The Policy Review Committee worked on long-term issues and was comprised of cabinet-rank officials from the relevant departments. However, due to the sensitive nature of the antisatellite issue an ad hoc Antisatellite Working Group made up of representatives from the State Department, DOD, CIA, JCS, and ACDA and chaired by Walter Slocombe (Principal Deputy Assistant Secretary for International Security Affairs) was set up. This group was separated from the main PRM-23 group to discuss ASAT-related issues.[84] The ASAT Working Group, as a result of President Carter's interest in antisatellite arms control, discussed U.S. negotiating strategy and its relationship to the U.S. ASAT program. None of the group wanted to curtail the program, yet neither did they support a crash program. Instead the group favored the development of the U.S. program in an orderly way that would facilitate the arms control process with the Soviet Union.[85] The group formulated the policy that the prospect of a U.S. ASAT capability would provide an incentive for the Soviet Union to negotiate, and would provide leverage during the negotiations. In addition, if an acceptable agreement proved elusive, the United States would have an ASAT capability.

The question of what form of limitations the United States should pursue caused the most disagreement within the Working Group. The Department of Defense initially favored the complete dismantling of the Soviet ASAT system, but as a result of either growing skepticism about the verifiability of a comprehensive ban, or a desire to maintain some ASAT capability for the United States it favored reaching a "rules of the road" agreement that would ban hostile acts in space.[86] However, the State Department and ACDA were more optimistic of a ban on testing and deployment. On September 3, 1977 the ASAT Working Group presented a range of arms control options to the President and on 23 September President Carter indicated his preference for comprehensive limits in the PRM/NSC-23 Decision Paper. The new Director of Defense Research and Engineering, William Perry, summarized the directive at the defense budget hearings in 1977:

> The PRM/NSC-23 Decision Paper dated September 23, 1977, requires that we seek a comprehensive ASAT agreement prohibiting testing in

space, deployment and use of AST capability . . . To reduce the possibility of a future space conflict, the President has directed that we seek an effective and adequately verifiable ban on antisatellite systems with the Soviets. As a consequence of this decision an interagency group—of which DOD is a part—has been making the necessary preparations for negotiating with the Soviets.[87]

The Presidential Directive's national security components remain classified, although the press release gave some indication of what had been decided:

The United States finds itself under increasing pressure to field an antisatellite capability of its own in response to Soviet activities in this area. By exercising mutual restraint, the United States and the Soviet Union have an opportunity at this early juncture to stop an unhealthy arms competition in space before the competition develops a momentum of its own. The two countries have commenced bilateral discussions on limiting certain activities directed against space objects, which we anticipate will be consistent with the overall U.S. goal of maintaining any nation's right of passage through and operations in space without interference. While the United States seeks verifiable comprehensive limits on antisatellite capabilities, in the absence of such an agreement, the United States will vigorously pursue development of its own capabilities. The U.S. space defense program shall include an integrated attack warning, notification, verification and contingency reaction capability which can effectively detect and react to threats to U.S. Space Systems.[88]

The press release was an offer of further U.S. ASAT restraint in return for reciprocal action from the Soviet Union. There was also a threat of a U.S. space defense program if the Soviet Union failed to conform.

Once President Carter had expressed his preference for comprehensive limits on antisatellites with PRM/NSC-23 the NSC began preparations for the negotiations with the Soviets. An Antisatellite Negotiating Working Group was established, although the departments represented were the same as the previous ASAT Working Group and as a result its membership was virtually identical. However, although PRM/NSC-23 had called for comprehensive limits on ASAT testing and deployment it had not specified how this was going to be achieved. The most important problem that arose was what activities and devices were to be prohibited. There was a large grey area over what systems constituted an antisatellite weapon. Electronic jammers and dual capable systems such as ABMs and ICBMs could all be used as a potential ASAT weapons. The Soviet Galosh exoatmospheric ABM system had a rudimentary ASAT capability.[89] A further important issue was how could a treaty prohibiting antisatellite weapons be verified. As the discussions continued the group became divided. The Defense Department became convinced that a comprehensive agreement would not be possible nor desirable. The principal coalitions were ACDA and the State Department favoring a comprehensive prohibition and the Defense Department and the Joint Chiefs of Staff against such an agreement.[90]

The Defense Department believed that the Soviets' dedicated ASAT weapon would be impossible to verify since the SS-9 booster was used for other missions, notably for launching ocean reconnaissance satellites. The argument that the Soviets would not have confidence in using a covertly deployed ASAT system because it had not been tested was countered by the fact that there were ways to disguise an ASAT test under the cover of activities such as spacecraft docking.[91] The State Department and ACDA on the other hand believed that the benefits of reaching an agreement which would curb antisatellite systems outweighed the risks of covert Soviet ASAT deployments. However, by March 1978, President Carter appeared to have become impatient with the negotiating group's division and decided to initiate formal discussions with the Soviet Union.

On 8 June of that year, talks began in Helsinki in the search for a comprehensive ban on ASAT weapons. During the first round of talks the U.S. delegation was headed by ACDA director Paul Warnke, whose position was to explore the extent of Soviet interest and thinking on the issue of ASAT arms control. The United States began the negotiations by proposing a complete prohibition of antisatellite weapons. However, it appeared that the Soviet delegation had not given serious thought to antisatellite arms control prior to the talks and subsequently asked for more time to consult with Moscow.[92] In addition to the prohibition of antisatellite weapons, the U.S. delegation explored various interim agreements which included a moratorium on the testing of antisatellite systems and a "noninterference" agreement. The Soviet delegation, headed by Oleg Khlestov, Head of the Treaty and Legal Affairs division of the Foreign Ministry wanted a guarantee from the United States that the space shuttle would not be used as an antisatellite weapon. However, the United States delegation had been ordered in advance to keep the space shuttle as a nonnegotiable subject.[93]

The next set of talks began on 16 January 1979 in Berne. The United States delegation sought from the Soviet delegation the range of possible agreements. It became apparent that the Soviets were willing to discuss a moratorium on antisatellite testing. They were not prepared to discuss the dismantlement of their antisatellite system. This position would have left the Soviets with their antisatellite capability intact while the United States would not have been permitted to develop its own system on a par with the Soviet system. It was therefore unacceptable to the United States' delegation. The only common ground was a nonuse agreement.[94] However, there were problems with this too. The Soviet position was that the nonuse would apply only to U.S. and Soviet satellites leaving allied satellites which were vital to NATO vulnerable. The Soviets also persisted with their objections to the space shuttle.

The third set of talks began on 23 April 1979 in Vienna and subsequently turned out to be the final talks. The combination of the Soviets" unwillingness to dismantle their satellite interceptor and the Defense Department's opposition to a comprehensive agreement, led the U.S. delegation to compose a two-stage strategy. The United States would seek a no use agreement possibly combined with a moratorium on antisatellite testing in the short term to be followed in the long run with an agreement to prohibit the hardware.[95] Progress was made during the talks of a no use treaty and a test moratorium was discussed, with the United States in favor of a short term moratorium and the Soviet Union in favor of a longer term one for reasons discussed above. However, further progress was prevented in the redrafting of a treaty regarding no use by the Soviets" repeated objections to the space shuttle and their desire to restrict a no use agreement to only U.S. and Soviet satellites. In addition to this, the Soviets also reserved the right to circumvent a "no use" agreement if "hostile or pernicious" acts by a foreign satellite infringed their national sovereignty. This was interpreted as a Soviet wish to prevent the potential use of direct broadcasting satellites for propaganda purposes.[96]

This round of talks was adjourned with the prospect of a fourth to be held in the autumn. However, further talks were delayed by the pursuit of SALT II discussions which were occurring and became the overriding priority of the Carter administration. Some officials in the Carter administration felt that the Joint Chiefs of Staff's support for SALT II might be at stake if they pushed too hard for an antisatellite agreement. The ensuing delay to the negotiations and the Soviet invasion of Afghanistan in December 1979 meant that any notion of an ASAT agreement was thwarted. There were a number of attempts within the administration in 1980 to reconvene the negotiations, mainly from within the State Department and ACDA and the Soviets also informally showed some interest in the resumption of talks. However, with Soviet ASAT testing resuming on 3 April 1980 the Carter administration gave up any hope of an agreement before the U.S. presidential elections in that year. The subsequent election of President Reagan brought U.S. interest in an ASAT arms control treaty to an end.

The U.S. Antisatellite Program since the End of the Cold War

The demise of the Soviet Union did not seem to have a corresponding effect on antisatellite proposals in the United States. Proposals in the early 1990s were argued from the Cold War premises which had dominated the debate since the issue arose in the Kennedy administration. These premises were founded on the 'space as a sanctuary" argument, that antisatellites would undermine strategic stability and the argument that an arms race in space would occur. The flaw in these premises was that the arms race in space issue combined with strategic stability were inextricably linked with a fully fledged adversary to maintain this relationship; with the disintegration of the Soviet Union this relationship disappeared. The Gulf War in 1990-1991 also changed the equation with the realization of the importance of space in warfighting during the campaign, and the possible effects if these assets were threatened.

The strategic arguments that had dominated the earlier ASAT debates began to dissipate in the post-Cold War and post-Gulf War, except in the purist position of supporting space as a sanctuary. However they were replaced by other concerns. The Clinton administration let its opposition to military space programs be known both in words and actions.[97] ASAT proponents, including retired Air Force General Charles Horner and the Secretary of the Air Force, attempted to raise the ASAT issue as one that needed addressing.[98] This was met with considerable opposition from the Clinton administration which followed the space sanctuary view and did not support the notion of a requirement for ASAT weapons. In particular, the Clinton administration resisted the Army's Kinetic Energy Antisatellite (KE-ASAT).[99]

However strong Congressional support, through Senator Bob Smith, has backed military space programs and kept some of them in existence in the face of opposition from both the Clinton administration and intraservice ambivalence. For example, Congress approved $30 million in 1996 for funds for KE-ASAT but this funding was rescinded by President Clinton. Congress rejected that action on 9 June 1996 by witholding money from some of the administration's favored projects.[100] However with the FY1998 budget Clinton vetoed specific programs from the budget which included the KE-ASAT, and cut $38 million from the project. Congress however managed to keep the KE-ASAT program going and in May 1998, KE-ASAT scientist Mark Fisher stated that "If there's money available we could conduct a proof of principle flight within 18 months. I would need $65 million to do two flight tests."[101] There were funds available in 1999 but this was from the previous year's funding. A request for an infusion of $41 million for the FY00 budget was requested to keep the program going although this was subject to political opposition.[102]

The Mid-Infrared Advanced Chemical Laser (MIRACL) in existence at White Sands testing range in New Mexico is perhaps closer to deployment than the KE-ASAT. It was originally an SDI antimissile program, but is in the process of being adapted into a laser for use against satellites. In addition to MIRACL the Pentagon is working on both excimer and free-electron lasers as ground-based ASAT systems.[103] These directed energy systems are able to respond in a more timely manner than kinetic energy systems. On 17 October 1997 the U.S. Army Space and Missile Defense Command used the Mid-Infrared Advanced Chemical Laser (MIRACL) ground-based laser to illuminate the MSTI-3 (Miniature Sensor Technology Integration) satellite in what was a test of satellite vulnerability.[104] The target satellite the MSTI-3 which was in a 265-mile circular polar orbit carried a mid-infrared, near-infrared, and visual focal plane array with a telescope and had finished its intended mission. The MIRACL used various power levels on the target satellite when it was 60-70 degrees above the horizon. The satellite and its sensors were not damaged since the intention was to test the level at which the laser caused the degradation of the sensor.[105] The MIRACL used excited deuterium fluorine molecules to produce 3.8 micron wavelength light for good atmospheric transmission. The power output is around 2 megawatts. The beam was aimed by the Hughes Sea Lite Beam Director.[106]

Defense Department Officials have been reluctant to provide information regarding what was obtained from the test firing, which included lasings by MIRACL and a low-power chemical laser.[107] The test cost about $2 million, with MIRACL operations running about $6,000 per second. Two bursts from the laser struck a sensor array on the MSTI-3 satellite. One burst was an initial one second firing to calibrate the laser's location on the satellite's body. The second beam was a 10 second burst, which triggered the sensors and relayed data back to the ground tracking and monitoring stations.[108] The Army experienced problems with the test which curtailed its effectiveness. Telemetry from MSTI-3 that was supposed to provide information about the test was never received. This information was to identify the power levels at which the MSTI sensors were blinded. The aim was not to damage the sensors but to temporarily blind them. A further problem that occurred during the testing was a shockwave in the laser cavity that damaged the MIRACL during its operation.[109] The satellite lasing was one of several initiatives the Army was considering in 1997. It is also conducting simulations to determine the effectiveness of other antisatellite weapons.

The MIRACL testing became the fulcrum of the debate on the issue of space control. The goal of space control has long been a part of the United States National Military Strategy, however, that did not include an antisatellite capability. The Pentagon under the Clinton administration did not consider the development of antisatellite weapons a priority. This test therefore was extremely controversial. The test was designed to measure the vulnerability of U.S. satellites to laser attack, but at the same time it was able to measure MIRACL's potential use as an emergency antisatellite weapon. Indeed, it has been claimed that MIRACL and its associated beam director has had an ASAT mission since the mid-1980s and has had a contingency mission to negate satellites harmful to U.S. forces.[110]

The testing which followed the lasing of the satellite has been focused on the Sea Lite Beam Director (SLBD) which is designed to track targets and help the laser beam target them. The SLBD was originally designed to track tactical targets such as aircraft and missiles, but improvements have increased the beam director's accuracy to enable it to track space objects. Although, the testing has not involved firing the MIRACL laser it has been oriented toward improving the overall system performance and operability. Indeed recent tests have been conducted on the task of keeping a laser focused on a target in space. In order to solve this problem, a target with the correct type of reflectors was required to gauge the effectiveness of the tests. The only available target with the required infrared "retroreflectors" is a U.S. satellite, the Low-Power Atmospheric Compensation Experiment (LACE). This is a "dead" satellite which enables the beam to track it.[111]

The beam director uses a technique known as a 2-D conical scan (Conscan) which involves moving the laser beam in a circle until it reflects light off the target back to the ground. Once the target is located, the beam director's boresight loop enables operators to keep the laser pinpointed on the target for as long as required. The Conscan experiments are designed to verify that the beam director can assist with reliable initial positioning and maintenance of a focused spot on an object. On 24 March 1998 the Army conducted its first active control scan boresight corrections using a satellite target which involved the beam director and the Low-Power Chemical Laser, a satellite tracking beam. This was referred to as the Data Collection Experiment (FY98 DCE).[112] The beam director tracked the LACE satellite at a distance of 550 kilometers and propagated the low-power laser beam (around 32 watts). The return energy was detected above 50 degrees in elevation, the lower elevation limit of the return off the LACE corner cube reflectors, and a Conscan track loop was closed.

A U.S. Defense Department directive (DOD I 3100.11) is the driving force behind the world's satellites being evaluated for their vulnerability to lasers. The work is being undertaken by the Satellite Assessment Center of the Air Force Research Laboratory's Directed Energy Directorate. The work is being undertaken in response to the new Defense directive which reflects two factors: there is an increasing number of satellites in space and some of these are particularly vulnerable to laser radiation.[113] The Satellite Assessment Center compiles detailed satellite intelligence coupled with laser effects testing on actual spacecraft components and materials to build high-fidelity computer models of foreign and domestic satellites. Using these models, the safe levels of laser illumination for a particular satellite are determined.[114] Another factor which is measured is the operation and orientation of particular satellites in relation to the proposed laser scenario. To help minimize costs the center is developing software upgrades that will provide U.S. Space Command the ability to screen satellites in-house. This software includes a centrally developed satellite vulnerability database that can perform predictive avoidance analysis as situations arise.

A further potential antisatellite system is the Airborne Laser. This is a modified Boeing 747 that will have the ability to fire directed energy at potential targets. A latent capability exists in using the Airborne Laser as an antisatellite weapon. The primary problem in using the ABL as an ASAT weapon arises from the use of infrared technology to track targets and cue the laser.[115] This requires a bright infrared reflection from the target. To use the ABL in an ASAT mission role an active system such as radar would have to be used to detect the satellites. There is also a difference of opinion whether the deconfliction system in the development for the ABL would be able to be incorporated in an ASAT role. The deconfliction process is used to ensure that the long-range radar does not intentionally hit an aircraft or satellite in front of or behind the target. At present Pentagon officials are not interested in developing the technology required for the ABL to be able to operate as an ASAT weapon. However, Air Force and aerospace industry officials believe that in the future the ABL may be given the task of intercepting satellites within 200 miles of the Earth's surface. It can be assumed that the ABL could destroy most low-Earth orbit satellites given its ability to deploy to a precise location that the satellites must fly over. The ABL is seen as a competitor to the congressionally supported Army program that is developing a ground-based ASAT capability designed upon an advanced kinetic-kill vehicle.[116]

The internal prioritization given to these programs was initially not high. The particular importance given by the military's policy and programs can be determined by whether or not they appear in the Five Year

Development Plan (FYDP) from which the services plan and what organization takes the lead. When funds are unrequested by DOD or the individual services it can be assumed that the programs are rogues rather than mainstream priorities. In this instance the programs were advocated from Congress, in particular the Senate, rather than the services themselves.[117]

The rationale behind the acquisition of a U.S. ASAT weapon system is that obtaining a proved means of disabling a satellite will discourage other countries from relying on them too heavily. The testing of an ASAT system would allow the military the confidence that it would be able to control the use made of space by future adversaries. It is this argument that weighs heavily in the thinking of the U.S. military and is vital for its future military operations.

Russian ASAT Activities since the End of the Cold War

The attitude to ASAT weapons in Russia has often been categorized as a response to U.S. ASAT developments combined with a response to U.S. ballistic missile defense efforts. Whether this is just rhetoric or is part of a wider development for Russia to continue to develop its ASAT capabilities is hard to determine. What can be said with some certainty is that Russia has carried out some activities with regard to ASAT weapons as the following demonstrates.

There have been reports of testing in Russia of a high altitude weapon which fired off an electromagnetic pulse or EMP that is similar to bursts caused by nuclear blasts.[118] This has the ability to disrupt a satellites' functioning. This test was seen as part of Russia's efforts to improve its antisatellite weapons technology. However due to the indiscriminate nature of EMP, directional weapons need to be used. Other activities in Russia are related to reports of a Russian air launched ASAT capability similar to the U.S.-developed F-15 miniature homing vehicle developed in the late 1980s. It was reported that a Russian Mikoyan MiG-31 was observed to be carrying an antisatellite weapon on its center under fuselage stores position.[119] The MiG-31 which was primarily designed as a high-altitude, long-range interceptor was modified slightly for its role as an ASAT carrier. Although these two reports do not constitute a concerted ASAT development plan, combined with the previous rigorous ASAT development they highlight the fact that Russia is still concerned with the issue of ASATs.

Conclusion

This chapter has highlighted the Soviet ASAT development of its co-orbital attack capability. The rigorous ASAT development and testing enabled the Soviet Union to have a reliable operational capability probable from the mid-1970s onwards. Indeed, the Soviets tested their ASAT capability on over twenty occasions against target satellites in varying orbits and inclinations and operated numerous attack profiles. There can be little doubt that the intended targets for this capability were U.S. and NATO satellites. It is possible that Russia could quickly operationalize the direct ascent co-orbital ASAT capability. It is however difficult to put a timeframe on how quickly they could operationally this. Since the latter stages of the Cold War and indeed since the collapse of the Soviet Union, Russia has shown some interest in an ASAT capability, of most note being the possible adaptation of the MiG-31 in an ASAT carrier role. Whether this interest will be developed upon is dependent upon the perceived threat Russia feels in response to the United States' missile defense and ASAT plans.

The different approach taken by the Soviet Union and the United States toward the development of an ASAT capability can be explained by timing and how seriously each country considered the issue. At the time of Soviets" interest in developing an ASAT the air-launched system was not technically viable. Also, the Soviets wanted a robust ASAT system which was provided by the coplanar space intercept. The U.S. on the other hand during the Cold War did not want to develop such a robust ASAT system, and merely wanted a limited system which the air-launched system provided.

The United States development of its ASAT capability saw the U.S. Army and Air Force compete for the mission. The initial U.S. ASAT policy utilized the ASAT Program 505, the NIKE-ZEUS anti-ballistic missile and was under Army command. However, when Program 505 was phased out and Program 437, which used the intermediate range ballistic missile the Thor, received the ASAT mission, the Air Force took the mission from the Army. This program itself was terminated and it was not until the conception of the air launched ASAT that the Air Force continued to hold the ASAT mission. Since the demise of the Soviet Union and subsequent end of the Cold War, the Army has been at the forefront of ASAT efforts, both in terms of the Ke-ASAT and the MIRACL testing.

Notes

1. William J. Perry, Brent Scowcroft, Joseph. S. Nye, Jr., and James A. Schear, "Antisatellite Weapons and U.S. Military Space Policy: An Introduction," in *Seeking Stability in Space: Antisatellite Weapons and the Evolving Space Regime* (Aspen: University Press of America, 1987), 1.
2. Paul B. Stares, "Antisatellite Arms Control in a Broader Security Perspective," in *Seeking Stability in Space: Antisatellite Weapons and the Evolving Space Regime,* 111.
3. Herbert York, *Race to Oblivion* (New York: Simon and Schuster, 1970), 131.
4. Quoted from U.S. Congress, Senate, Committee on Aeronautical and Space Sciences, *Soviet Space Programs 1962-1965: Goals and Purposes, Achievements, Plans and International Applications,* Staff Report (30 December 1966), 75.
5. U.S. Congress, Senate, p. 75.
6. Curtis Peebles, *Battle for Space* (Dorset: Blandford Press, 1983), 83.
7. Peebles, 85.
8. Peebles, 88.
9. Peebles, 94.
10. Donald L. Hafner, "Averting A Brobdingnagian Skeet Shoot," *International Security,* Winter 1980/81, 51.
11. Secretary of Defense Donald H. Rumsfeld, *Report of the Secretary of Defense Donald H. Rumsfeld to the Congress on the FY 1978 Budget,* (Washington, D.C.: U.S. GPO, 1979).
12. Secretary of Defense Harold Brown, *Department of Defense Annual Report FY 1979,* February 2, 1978.
13. Roger C. Hunter, *A United States Antisatellite Policy for A Multipolar World* (Maxwell Air Force Base, Alabama: Air University Press), October 1995, 19.
14. Marcia S. Smith, *CRS Issue Brief: ASATs— Antisatellite Weapon Systems,* Congressional Research Service, December 7, 1989, 15.
15. Smith, 15.
16. R. C. Hunter, *A United States Antisatellite Policy for A Multipolar World,* 20.
17. President Reagan, *U.S. Policy on ASAT Arms Control: Communication from the President of the United States* (Washington, D.C.: USGPO, 1984), 7.
18. Reagan, 8.
19. Reagan, 8.
20. Richard L. Garwin, Kurt Gottfried and Donald L. Hafner, "Antisatellite Weapons," *Scientific American,* June 1984, 31.
21. Curtis Peebles, *Battle for Space,* 115.
22. Peebles, 155.
23. Peebles, 116.
24. Peebles, 32.
25. Stephen Kirby et al., *The Militarization of Space* (Sussex: Lynne Riener, 1987), 110
26. Kirby, 110.
27. Kirby, 110.
28. Kirby, 110.
29. Kirby, 117.
30. Marcia S. Smith, *CRS Issue Brief: ASATs—Antisatellite Weapon Systems,* 12.

31. Office of Technology Assessment, *Antisatellite Weapons, Countermeasures, and Arms Control: Summary* (Washington, D.C.: U.S. Government Printing Office, 1985), 10.
32. R. C. Hunter, *A United States Antisatellite Policy for A Multipolar World*, 24.
33. Nicholas L. Johnson, *Soviet Military Strategy in Space* (London: Jane's Publishing Company, 1987), 139.
34. USAF Series on Soviet Military Thought, no. 9, *Dictionary of Basic Military Terms: A Soviet View* (Washington, D.C.: U.S. GPO, 1972), 177.
35. Curtis Peebles, *Battle for Space* (Dorset: Blandford Press, 1983), 102.
36. Clarence A. Robinson, Jr., "Antisatellite Weaponry and Possible Defense Technologies against Killer Satellites," in Uri Ra'anan and Robert L., Pfaltzgraff, *International Security Dimensions of Space* (Hamden, CT: Archon Books, 1984), 71.
37. Robinson, 71.
38. Paul B. Stares, *The Militarization of Space* (New York: Cornell University Press, 1985), 137.
39. G. E. Perry, "Russian Hunter-Killer Satellite Experiments," *Royal Air Force Quarterly*, vol 17 (Winter 1977), 329.
40. Charles S. Sheldon, 'soviet Military Space Activities" in *Soviet Space Programs 1971-75*, Committee on Aeronautical and Space Science, U.S. Senate, 1976, 426.
41. *Soviet Military Power* (Washington, D.C.: U.S. Department of Defense, March 1983).
42. Paul B. Stares, *The Militarization of Space*, 138.
43. Nicholas L. Johnson, *Soviet Military Strategy in Space*, 143.
44. Paul B. Stares, *The Militarization of Space*, 138.
45. Nicholas L. Johnson, *Soviet Military Strategy in Space*, 144.
46. Paul B. Stares, *The Militarization of Space*, 138.
47. G. E. Perry, "Russian Hunter-Killer Satellite Experiments," 332.
48. Paul B. Stares, *The Militarization of Space*, 139.
49. G. E. Perry, "Russian Hunter-Killer Satellite Experiments," 333.
50. Nicholas L. Johnson, *Soviet Military Strategy in Space*, 144.
51. Johnson, 146.
52. Johnson, 146.
53. Johnson, 146.
54. Johnson, 146.
55. Nicholas L. Johnson, *Soviet Military Strategy in Space*, 148.
56. Paul B. Stares, *The Militarization of Space*, 143.
57. Curtis Peebles, *Battle for Space*, 107.
58. Peebles, 109.
59. G. E. Perry, "Russian Hunter-Killer Satellite Experiments," 333.
60. Paul B. Stares, *The Militarization of Space*, 144.
61. S. M. Meyer "Soviet Military Programs and the New High Ground," *Survival*, vol. 25, no. 5 (September/October1983), 210.
62. Nicholas L. Johnson, *Soviet Military Strategy in Space*, 150.
63. Johnson, 150.
64. "Satellite Killers," *Aviation Week and Space Technology*, vol. 104 no. 25, (21 June 1976), 13.
65. Paul B. Stares, *The Militarization of Space*, 187.
66. Stares, 187.
67. Stares, 188.
68. Nicholas L. Johnson, *Soviet Military Strategy in Space*, 150.

69. Paul B. Stares, *The Militarization of Space*, 188.
70. Nicholas L. Johnson, *Soviet Military Strategy in Space*, 152.
71. Paul B. Stares, *The Militarization of Space*, 188.
72. Nicholas L. Johnson, *Soviet Military Strategy in Space*, 152.
73. Johnson, 152.
74. Johnson, 153.
75. Johnson, 153.
76. Secretary of Defense Harold Brown declared the Soviet ASAT was operation in 1977 along with the *Soviet Military Power,* 1981 (Washington, D.C.: U.S. Department of Defense, 1981) 68 and *Soviet Military Power,* 1983, 67-68. *The Soviet Military Power,* 1984, 24 and *Soviet Military Power,* 1985, 55 declared that the Soviet ASAT was operational in 1971.
77. Lawrence Freedman, "The Soviet Union and Anti-Space Defense," *Survival*, January/February 1977, 23.
78. Nicholas L. Johnson, *Soviet Military Strategy in Space*, 156.
79. Johnson, 156.
80. Johnson, 156.
81. Press Conference of President Carter, Washington, D.C., March 9, 1977, reproduced in Roger Labrie, ed., *SALT Handbook: Key Documents and Issues*, (Washington, D.C., American Enterprise Institute: 1979), 423.
82. Press Conference of Secretary of State Vance, Moscow, March 30, 1977 in Labrie, *SALT Handbook: Key Documents and Issues*, 429.
83. White House Press Release, *Description of a Presidential Directive on National Space Policy* (The White House, June 20, 1978).
84. Paul Stares, *The Militarization of Space*, 182.
85. Stares, 183.
86. Stares, 184.
87. U.S. Congress, House, Subcommittee of the Committee on Appropriations, *Department of Defense Appropriations for 1979*, Hearings, 95th Congress, 2nd Session (1978), Part 3, 726-7, quoted in Paul Stares, *The Militarization of Space*, 184.
88. White House Press Release, *Description of a Presidential Directive on National Space Policy* (The White House, June 20, 1978).
89. Paul Stares, *The Militarization of Space*, 194.
90. Stares, 194.
91. Stares, 195.
92. Stares, 196.
93. James Canan, *War in Space* (New York: Harper and Row Publishers, 1982), 24-25.
94. Paul Stares, *The Militarization of Space*, 198.
95. Stares, 198.
96. Stares, 199.
97. Joan Johnson-Freese, "The Viability of U.S. Antisatellite Policy: Moving Toward Space Control," *Institute for National Security Studies Occasional Paper 30*, January 2000, 24.
98. Steve Weber, "ASAT Proponents Fail to Reverse White House Policy," *Space News*, September 19-25, 1994, 7.
99. Pat Cooper, "U.S. Political Battles Threaten Antisatellite Project," *Space News*, June 17-24, 1996, 6.
100. Cooper, 6.

101. Jonathan Wright, "U.S. Military Moves Closer to Star Wars," *Daily Telegraph*, May 21, 1998, 4.

102. Joan Johnson-Freese, "The Viability of U.S. Antisatellite Policy: Moving Toward Space Control," 25-26.

103. Joan Johnson-Freese, 26.

104. Michael A. Dornheim, "Laser Engages Satellite, With Questionable Results," *Aviation Week and Space Technology*, October 27, 1997.

105. Dornheim.

106. Dornheim.

107. Robert Wall, "Army Reconsiders Lasing Satellite," *Aviation Week and Space Technology*, December 20, 1999.

108. Frank Sietzen, "Laser Hits Orbiting Satellite in Beam Test," *SpaceDaily*, October 20, 1997 http://www.spacedaily.com/news/laser-97a.html.

109. Robert Wall, "Army Reconsiders Lasing Satellite."

110. 'service More Tight-Lipped About MIRACL's ASAT Mission," *Inside the Army*, November 30, 1998 Vol. 10 No. 47, 2.

111. *Inside The Army*, 2.

112. *Inside The Army*, 2

113. "Lab Evaluates Satellites," *U.S. Air Force Research Laboratory*, July 6, 2000, Office of Public Affairs, http://www.de.afrl.af.mil/News2000/00-50.html.

114. "Air Force Lab Evaluating Satellites Vulnerability To Lasers," *SpaceDaily*, July 12, 2000 http://www.spacedaily.com/news/laser-00i.html.

115. David A. Fulghum, "Laser Offers Defense Against Satellites," *Aviation Week and Space Technology*, October 7, 1996, 27.

116. Fulghum, 27.

117. Fulghum, 27.

118. "Russian ASAT," *Washington Times*, June 18, 1999, 9.

119. "Russians Alter MiG-31 For ASAT Carrier Role," *Aviation Week and Space Technology*, August 17, 1992, 63.

Chapter Six

The Space-based Laser for Ballistic Missile Defense

This chapter addresses the issue of weapons actually being based in space. At present the principal weapon under consideration is the space-based laser. The chapter initially explores the scientific basis for space-based lasers including the chemical reaction which is required to produce the laser beam. The importance of the science and technology aspect of a laser weapon cannot be overstated. The focus here is on chemical laser weapons, rather than free electron lasers or nuclear-pumped lasers, because the science and technology of these other types of lasers during the SDI research and development period showed them to be unfeasible. Also, the science and technological aspects of laser weapons became important politically, when the Union of Concerned Scientists entered the SDI debate. It then became important to the policy debate to become scientifically informed. The lethality of a space-based laser is then considered with particular relation to the distance and intensity of the beam required to intercept ballistic missiles in flight. The orbital characteristics of a space-based laser are analyzed, and the basing of the satellites is addressed in relation to the orbit type and inclination required. The components that would be required to operate a space-based laser are addressed in the following section. The final sections outline the technological programs that comprise the space-based laser. It explores industry's involvement with the U.S. Air Force and the Ballistic Missile Defense Organization in the development of the space-based laser, and outlines the timetable for the eventual deployment of the space-based laser.

Interest in utilizing space-based lasers for ballistic missile defense arose when two facts emerged. One was that ballistic missiles are relatively fragile. They do not resist laser energy. The second fact was that chemical lasers could project missile killing amounts of energy over 3,000 kilometers. These two facts more than anything else saw political interest soar over the possibility of placing laser weapons in space. Laser weapons located in space could be used to intercept ballistic missiles in their boost phase.

The Science and Technology of Laser Weapons in Space

Light is generated from lasers by the phenomenon known as Light Amplification by the Stimulated Emission of Radiation (LASER). Pumping energy into molecules causes them to be in an excited electronic state, in which they possess an amount of energy. The molecules then release their energy as photons. These photons when passing by an excited molecule cause that molecule to release its energy as another photon. This process continues to emit further photons. To achieve great light amplification in the laser, the beam has to pass through a large amount of laser medium. To create this effect, the light beam is passed repeatedly through the same laser medium by reflection back and forth between mirrors at the opposite ends of the laser recess. The beam leaves the recess through one of the mirrors which is partially transparent.

To create a continuous laser beam, energy has to be continuously pumped into the laser medium. The conversion of fuel energy into excited energy and subsequently into laser photons is not very efficient. Energy which is not emitted as photons in the laser beam is removed from the system as heat. In a chemical laser gas dynamic expansion is used for pumping and cooling, and electric discharge pumping is sometimes used to supply the required amount of activation energy in order for the exothermic reaction to produce the excited lasant. While a fraction of the energy increases the kinetic energy of the molecules and heats them, some of the energy is absorbed into the internal vibration and rotational motions of the molecules. This results in the low energy states being depopulated and a significant number of molecules entering an excited state. This condition is known as population inversion.[1] The rate at which a laser can emit photons is firstly dependent upon the rate at which the electrons from the ground state can be pumped to the excited state. Secondly, it is dependent on the rate at which the intermediate energy level is made available to electrons to cascade from the excited state, since the exclusion principle does not permit more than one electron to occupy a given state of an atom.[2]

In a chemical laser the total energy output, the total latent heat of combustion, and the total electric emission activation energy required are proportional to the number of lasant molecules produced. In comparing the performance of chemical lasers, both the electrical efficiency (laser output energy divided by the electric emission energy) and the chemical efficiency (laser output energy divided by the latent heat energy of combustion) are significant measures of performance.[3] The specific power (the laser output energy divided by the mass of the reactants consumed) is more relevant than the chemical efficiency. Some chemical lasers do not require any electrical discharge for their operation. In chain reaction lasers the reaction products which are produced early in the reaction catalyze the rest of the reaction, which means that only a small initial electric charge is required. Electrical efficiency in these types of lasers is an insignificant criteria for performance.

The amount of energy focusing on a target that is formed by a distant laser is dependent on not only the intensity of the light source but on the fate of the laser beam as it travels through the space occupied between the laser and the target. When the laser is based in space the beam suffers from more diffraction. Diffraction is a consequence of the wave nature of light and occurs wherever it is based. Every point of an aperture can be treated as a point source of light with an appropriate phase and amplitude, and the light field at a distant point is the sum of the light fields of the spherical waves radiated by the light points at the aperture.[4] If the phase and amplitude are uniform over a circular aperture of diameter, the field divergence of the laser beam will have the minimum diffraction limiting angle. However, in practice the amplitude and phase of the laser beam will vary across the laser aperture. The formation of multiple traverse laser modes, or imperfection in the optics, or surface quality of the focusing mirror will make the diffraction angle greater than the minimum diffraction limiting angle.[5]

In a chemical laser the two elements or chemical compounds are linked to form molecules of a new compound. The mixing and ignition of two fuels (for example, hydrogen and fluorine, or oxygen and iodine) in a combustion chamber creates the power source for a laser. In order to produce a laser beam, deuterium, nitrogen trifluroide, and helium are mixed to produce fluorine with hydrogen in a mirrored chamber called an optical resonator. The compound, with its outer electrons which are in orbit barely attached, exits the chamber at hypersonic speeds through special nozzles. As it cools significantly its outer electrons snap back releasing energy in the form of light. The molecules are created in an excited state. It is possible by controlling their environment to achieve simulated emission of radiation before they return to their ground state by dissipating their energy as heat. An optical resonator amplifies the cascade of photons, transforming them into a laser beam. The power of the chemical laser is dependent on the number of molecules that can be lased at any particular time. In order to increase the power more fuel has to be burned and more nozzles need to be created. The efficiency of the laser is dependent on the speed at which the hot molecules exit. Subsequently, the purer the vacuum that draws out the molecules, the better the laser operates. In ground-based lasers, the vacuum is produced by steam pumps or by chemical absorption. In space, chemical lasers run at top efficiency due to the nature of space.

In a hydrogen-fluorine (HF) chemical laser, atoms of hydrogen combine with those of fluorine to produce an excited molecule of hydrogen-fluorine. This hydrogen-fluorine molecule emits at a wavelength of 2.7μm. An advantage of a chemical laser is its potential for relatively high conversion efficiencies. The HF chemical laser is the leader in projected power output. In Project Alpha, TRW assembled a 2-3 Mega Watts (MW) HF chemical laser.[6] Project Alpha was set up to demonstrate that a HF chemical laser could be scaled up to a power output level of 5 to 10MW.

Another chemical laser, although unsuitable for scaling up was tested. This laser called the Mid-Infrared Advanced Chemical Laser (MIRACL) uses deuterium fluorine (DF) as laser material and can produce a power output in excess of 1MW. This laser was successfully ground-based tested against a second-stage of a Titan I missile at a range of 0.5 miles.[7] The DF laser operates in the same way as the HF laser, except that the hydrogen is replaced by the deuterium. The DF laser emits infrared light at a wavelength of 3.8 micro meters (μm). Air is transparent at this wavelength, although this is not the case at the HF wavelength of 2.7μm, which means that the DF laser is beneficial when the beam has to penetrate the atmosphere.

The Lethality of a Space-based Laser

The ability to deliver a high-intensity beam of light long enough to disable a target is the objective of a laser weapon. After generation in a laser chamber, laser light is focused on a target by a system of mirrors. In order to change the focus point of the laser the position of the mirrors has to be altered. To fire the laser beam in various directions, the mirrors need to be rotated. The laser energy can damage boosters if the laser has a moderate intensity combined with a sustained dwell period on the booster. The laser will then burn through the missile skin. A 10 meter mirror with a HF laser beam would yield a 0.32 microradian divergence angle and create a laser spot 1.3 meter in diameter at a range of 4,000 meters. The distribution of the 20MW over the laser spot would create an energy flux of 1.5 kilowatts per square centimeter (kW/cm^2). The laser spot would need to dwell on the target for 6.6 seconds to create the nominal lethal fluence of 10 kilojoules per square centimeter (kJ/cm^2).[8]

At a range of 2,000 meters the destruction of the booster would require 1.7 seconds of illumination.[9] The laser beam would have to lead the booster by around 50 meters in order to take account of the time taken for the laser to travel from source to the missile which would be in flight.

A solid fuelled booster could probably absorb without disruption approximately $10kJ/cm^2$ on its skin.[10] This energy fluence would be the result from a 1 second illumination at $10kW/cm^2$. The application of an ablative material would probably double or maybe even triple the lethal fluence required. It is argued that the use of a mirrored reflective coating to the booster would deflect the laser, but the abrasion during the boost phase could cause it to lose its reflective capabilities. Another method of countering lasers is spinning the booster which could increase its resistance by a factor of three. The spinning minimizes the damage because of the shorter delve time of the laser.[11] The spot where the laser hits the booster is illuminated for a shorter period as the booster is spinning. However, it is possible that the uniform heating around the circumference of the booster could introduce a lethal mechanism which destroys the booster.

The process of destruction by a laser weapon begins when the laser reaches the target and lasts as long as it is fixed on the target. The type of the laser determines the amount of energy transferred. Other factors include the duration of exposure to the laser beam, the target, and certain environmental factors. The crucial factor is the ability of the target to absorb or reflect energy.[12] If the target possesses a high reflectivity to the laser energy it is difficult, maybe even impossible to inflict damage. The main destructive capability of a laser is thermal energy. On the occasion when the illuminated surface of the target is incapable of reflecting or absorbing the energy safely, a rapid buildup of heat occurs. This in turn melts and boils or vaporizes the target. Prior to the target material reaching melting point it may be seriously weakened. In addition to this, destruction may be inflicted by indirect mechanical stress caused by the intense heat of the laser.[13] The combination of thermal and mechanical factors may create thermomechanical effects and create the worst damage. This creates a series of pressure waves through the material which results in it being torn apart. An advantage of a laser weapon for ballistic missile defense is that the beams have the ability to travel vast distances at the speed of light above the atmosphere.

Many target materials may be cracked or perforated with less energy than would be required to melt through them. If the structural components are already under stress, like the skin of a missile through boost phase, only an incremental amount of additional stress needs to be added by the laser in order to exceed the yield stress of the material.[14] Once a flaw has been produced the structural stresses could concentrate on this flaw and propagate it, creating very large cracks. The susceptibility to crack propagation depends upon the strength, toughness, and fatigue properties of the target material along with the operational stresses to which it is subjected.[15]

For targets that are in the atmosphere, a beam with an intensity of around 10 million watts per square centimeter would cause the air immediately in front of the target to ionize which would create a layer of plasma as the beam hits the surface. The plasma would absorb the energy of the laser beam and grow extremely hot (around 6,000 degrees Celsius). The plasma would distribute this energy in two ways, by emitting ultraviolet radiation and by expanding explosively. These mechanisms could increase the extent of the beam energy attached to the target to approximately 30 percent and reduce the amount of energy the laser would have to produce.[16]

A space-based laser with a 20MW HF with 10 meter mirrors would have a laser wavelength of 2.7μm, but this would be attenuated as it disseminates through the atmosphere, although most of the light would reach an altitude around 10 kilometers.[17] Penetration deeper than this would not be required since the laser would not be in a position to attack missiles in flight until they had reached this altitude. Also, clouds could obscure the booster below a ceiling of 10 kilometers.

Requirements for Several Laser Weapons

	ASAT Space	Ground-based BMD	Space-based BMD
Laser type	chem (HF)	chem (DF)	chem (HF)
Laser wavelength	2.7μm	3.8μm	2.7μm
Laser location	space	ground	space
Target distance	3,000km	10km	3,000km
Mirror diameter	4m	4m	10m
Laser output	2.5MW	2MW	20MW
Time/shot (at maximum range)	75 secs	75 secs	8 secs
Beam spread	0.8μrad	1.2μrad	0.33μrad
Beam size at target	2.5m	2.5m	1m
Incident energy for kill	56W/cm^2	56W/cm^2	2500W/cm^2

Atmospheric transmission	100%	50%	100%
Laser efficiency	20%	20%	20%
Fuel energy content	1.4MJ/kg	1.4MJ/kg	1.4MJ/kg
Fuel per shot	720kg	est. 720kg	560kg

Source: Adapted from Dietrich Schroeer, "Directed-Energy Weapons and Strategic Defense: A Primer," *Adelphi Papers 221*, (London: IISS, Summer 1987)

Ground-based lasers have a problem with the transmission of the beam to the target since the laser has to travel through the earth's atmosphere. Lights of certain wavelengths are absorbed through the earth's atmosphere. The HF laser which emits light at a wavelength of 2.7µm is mostly absorbed, whilst that from a DF laser with a wavelength of 3.8µm transmits well. Air within the atmosphere also can defocus a beam of light in various ways. Intense laser beams heat the air around them which results in "thermal blooming" which at levels greater that 1MWcm^2 causes the beam to broaden. Thermal blooming is caused by a succession of sparks or plasmas which are created by the beam's energy that heats the air within the beam. The small heated areas make it difficult for the energy to penetrate, and this consequently causes the beam to diverge or bloom, reducing its efficiency at long ranges.[18] This diffusion of the beam makes it increasingly difficult to inflict damage on the target. The thermal blooming effect becomes a further complicating factor when it is necessary to track a moving target and hence move the laser beam through the air. The effect of thermal blooming increases rapidly as the beam power increases. Thermal blooming can be reduced by increasing the size of the beam. This can be achieved by increasing the diameter of the mirror; however the mirror must be increased in proportion to average power.[19]

The main effects of the atmosphere upon the laser beam are absorption, scattering, turbulence, thermal blooming, and spark generation. The turbulence in the air may cause the beam to defocus and bend. If the range of the laser beam is long, the beam cross section may be caused to deviate considerably from its circular shape. The inhomogeneity of the atmosphere along the path of the beam could act as a lens to create hot spots in the beam. These hot spots are localized areas of the beam where the intensity of the beam is greater than the average due to the localized focusing.[20] If separate rays in a laser beam have optical paths which differ in length by half a wavelength, the waves propagating along those rays will interfere destructively at the target. The propagation of the laser beam through the air can create what is known as Raman scattering. This effect changes the original wavelength of the beam to another or several other wavelengths.[21] This process may repeat itself and can lead to the beam losing a great deal of its efficiency.

A further difficulty in transmitting a laser through the atmosphere is the risk of creating a plasma. Since light waves are forms of electromagnetic radiation they create a strong electric field along with them. At intensities of around 10 million watts per square centimeter, the value of each depends on the frequency, it creates a field so strong that it removes electrons from the atoms in the air, which ionizes the air and creates a plasma.[22] The plasma proceeds to absorb the beam and subsequently interrupts its transmission. This effect places an upper limit on the intensity of a laser beam that can transmit through the atmosphere.

The Orbital Characteristics of a Space-based Laser

Directed energy weapons would be located on satellites placed in low-earth orbits. The type of orbit would depend on the nature of the threat. The satellite's orbital altitude is an important factor since it must place the laser, as frequently as possible, in a position where it can destroy the largest number of missiles in their boost phase. The satellite needs to be at an altitude sufficient to enable it to intercept the farthest boosting missile it can see without focusing the beam so long that closer and more vulnerable missiles are missed. The optimal altitude depends upon the height at which the booster's engines stop firing and also on the capacity of the lasers and the hardness of the missiles. When the Soviet Union was considered to be the main threat, polar orbits were chosen since they provided good coverage of the northern latitudes. However, polar orbits concentrate space-based lasers at the poles where there are no ballistic missiles deployed. The optimum configuration would be a number of orbital planes inclined about 70° to the equator.[23]

The Components of a Space-based Laser System

The satellites would not only contain the laser weapons, but equipment to perform surveillance, acquisition, tracking, damage assessment, and management functions. In boost phase defense, surveillance, acquisition, tracking, and kill assessments may be comparatively easy. The booster rockets have a hot plume making them easy to detect. Nevertheless, the booster's position must be located with high precision which means that a ranging method using visible light, a technique named laser-based radar (lidar) becomes necessary.[24]

To be able to aim a directed-energy weapon with enough precision requires the location and tracking of a ballistic missile booster. Boosters emit hundreds of kilowatts of power at short- and medium-wave infrared wavelengths of a few microns which sensors can detect. However, to be able to be beneficial for laser weapons to intercept ballistic missiles the sensors must be able to identify the booster within an area as small as the beam spot.[25] The beam would otherwise have to sweep across over an area on the missile not maximizing the laser energy's output and kill capability. Sensors with small angular resolution are required to interact with small divergence beams. A laser weapon ballistic missile defense system would have to be able to detect approximately a thousand targets, calculate the coordinates of each one continuously and hand over each one to the laser aiming and tracking system. This must be carried out quickly so that the laser can fire at each one within a few hundred seconds, although the number of targets depends on the capabilities of the potential enemy.

A HF space-based laser would have a spot size of 1.5m and 0.6m for a ground-based laser. A large infrared telescope with a 5m diameter observing booster emissions at a wavelength of around 4 microns would have an angular resolution of no more than a microradian. This would locate ascending boosters to within a spot 5m wide at 5Mm range.[26] This would therefore be inadequate for directing the laser beams at a point source. The targeting would be more demanding since the booster is not the point source. The booster plume is larger than the laser beam spot, so the booster body would have to be located in relation to the plume to avoid the beam attacking the plume. As mentioned above a laser radar (lidar) is required.

In order to achieve a finer angular resolution a shorter wavelength is required, in the visible or ultraviolet. At these wavelengths the sensor must provide its own illumination, hence the laser radar. A lidar operates by shining a low-power visible or ultraviolet laser on the booster body, and a telescope senses the reflected light.

The directed-energy beam needs to be aimed and stabilized. If the beam fluctuates, the effective divergence increases, and the beam misses the target. The mirrors directing the beam need to be stabilized despite vibrations caused by the beam's large power source. In the 15 milliseconds the beam requires to travel from source to the booster, the booster moves approximately 50 meters. The beam must lead the target. In one second of dwell time the target moves several kilometers. The beam must therefore remain on target whilst maintaining its aim and jitter control.

A further requirement of a space-based laser system, although not essential is kill assessment. The confirmation of an intercept would allow the beam to progress to the subsequent impending missiles. The structural damage to the booster would be evidenced by an erratic course or burn pattern. The damage to a bus would be hard to determine if the debris, including the reentry vehicles continued on their course.[27] In relation to kill assessment it is the assessment of whether the laser beam is missing the target, maybe by misalignment of the sensor or beam boresights, or miscalibration of aiming mechanisms that is important. If the beam is off target it is important to ascertain by how much and in what direction.

Decoy discrimination during boost phase is made relatively easy due to the fact the decoys must simulate both the energy output and the flight characteristics of the missile. Post-boost phase, the payload (the bus) of the missile continues to follow a ballistic trajectory. During the bussing phase interception is made difficult since relatively little energy is emitted by the bus, since only very small rockets are used for maneuvering. This makes the bus and the warheads less visible to infrared sensors. Optical sensors might track the target at this juncture and perhaps by visible light using lidar. Once the warheads have separated from the bus, from apogee, the slowest point in their freefall trajectory they gain speed. The reentry vehicles (warheads) are more resistant to damage from directed energy weapons than boosters and they also may have decoys surrounding them.

It is generally accepted that space-based laser weapons would be incapable of lasing the reentry vehicle with a sufficient dose of energy during its midcourse and reentry trajectory. The reasoning behind this lies in the fact that the reentry vehicle is hardened since it is designed to survive the launch phase, midcourse, and thermal reentry and successfully detonate and destroy even hard targets.[28] The booster must therefore be targeted during the time when it is above the clouds and atmosphere and the start of the deployment of the reentry vehicles by the bus. The access time to destroy the booster is approximately 100 seconds less than the time a typical ICBM would require to complete its launch phase prior to reentry vehicle deployment. Therefore the time the space-based laser would have to destroy the attacking booster is around 200 seconds.[29] Using the information from the space-based boost phase defense system table it identifies that the time required to destroy a ballistic missile is 1.1 seconds. It can therefore be estimated how many missiles a space-based laser can intercept. Assuming that around the same time would be required to retarget the laser to another booster, the laser could destroy around 100 boosters in the 200 seconds time that is available for interception.[30] That figure is the lowest possible number since it assumes that the 200 missiles are all launched simultaneously, and that none of the missiles launches are staggered at all. Also, if a space-based laser runs out of chemical ammunition, it is anticipated that it could be resupplied by a space plane or a remotely controlled space vehicle.

A Space-based Laser Boost-phase Defense System

Laser type	HF
Laser power	20 MW
Laser focusing mirror	10m diameter
Altitude of laser station	300km
Number of laser stations	120
Spacing between stations	3,600km
Average range for laser shots	1,300km
Number of shots per station	400
Average fuel per shot	82kg
Average time per shot	1.1 seconds
Total shots available (120x400)	48,000
Total fuel (48000x82kg)	3,900 tons
Total station mass (2 x fuel mass)	7,800 tons

Source: Dietrich Schroeer, "Directed-Energy Weapons and Strategic Defense: A Primer," *Adelphi Papers 221*, (London: IISS, Summer 1987)

The table outlines a set of performance parameters for a complete boost-phase strategic defense system. The system is assumed to have a boost-phase vulnerability period of 150 seconds, and a missile hardness of $20kJcm^2$. The target is assumed to be similar to the U.S. MX missile.[31]

A space-based laser is able to use the time it is in view of the missiles far more efficiently than kinetic kill vehicles based in space. A kinetic kill vehicle based in space may contain around twenty interceptors. The amount of missiles that a space-based laser could account for would equal the number of seconds that the satellite was in view of the missile, divided by the number of seconds it took to kill each. This means that potentially the number of kills per space-based laser could rise into the hundreds.[32] The laser weapon uses a relatively small amount of fuel to generate the beam and hence fire a shot. The potential is there to store a large number of shots in each laser system, and since the direction of the beam is determined by mirrors the laser has the ability to move from target to target over a wide field of view. Subsequently, a laser weapon has the ability to deal with a large number of targets, even if the targets are from a wide range of field.[33]

An advantage with reference to the ballistic missile defense mission of directed energy weapons over conventional missiles with high explosive warheads, is that destructive amounts of energy can be transmitted to the target at the speed of light, since no bulky container or explosive charge is involved in the process. Also, at present only a laser weapon has the capability to intercept an intercontinental ballistic missile during the boost phase of its flight. One disadvantage of laser weapons over conventional interceptors is that the beam must hit the target, which at long range raises serious target acquisition and tracking problems. With a conventional warhead a kill could occur if the warhead blast is sufficiently close to the missile in order to render it disabled.

One issue for space-based lasers is that of the vulnerability of the satellites. That is, how susceptible are the laser satellites to a direct attack on themselves? One way to defend the laser satellites is to hide the location of the satellites by producing small inexpensive inflatable tin foil copies of the satellite. Some of the decoy satellites could even be fitted with heat sources to simulate the heat put out by the real satellite.[34] The real laser satellites could also alter their radar signature, alter their thermal signature, and maneuver to another location. The concept of stealth could also be applied to satellites in orbit. Also, the laser satellites would have the option to fire at the interceptor that appears to be targeted towards it. The laser weapons could defend themselves against conventional nonnuclear interceptors and nuclear warheads.

The Origins of Industry's Involvement in Space-based Lasers

In the late 1960s efforts to develop and field high-energy laser weapons were initiated. The gas dynamic carbon dioxide (CO_2) laser was one of the earliest promising laser concepts developed in the United States by AVCO Everett in 1968.[35] This was shortly followed by the hydrogen fluorine (HF) and deuterium fluorine (DF) chemical lasers developed in 1969 by the United Technology Research Center (UTRC).[36] In the late 1970s the three U.S. military services in the 1970s started work investigating the military potential of laser weapons.

The United States Navy in 1978 performed a number of tests under the Unified Navy Field Test Program at San Juan Capistrano in California, in which a chemical DF laser around 400kW destroyed some TOW wire-guided antitank missiles in flight.[37] The laser was directed to the target by a Hughes aircraft aiming and tracking system. In 1980, a UH-1 helicopter was destroyed by the laser system.

The United States Air Force in 1981 using a laser aboard a Boeing NKC-135 cargo aircraft, called the Airborne Laser Laboratory, attempted and failed to shoot down an air-to-air AIM-9L Sidewinder missile while airborne. The testing continued, and in May 1983 the 400kW laser shot down a number of Sidewinder missiles.[38] The program was cancelled in 1984.

In 1984, the United States Army designed a small, compact DF laser called the Mobile Army Demonstration (MAD). The MAD was used as a prototype for an air defense weapon against missiles. It began at a power level of 100kW but was scaled up to 1.4MW.[39] The use of a DF laser caused the problem of poisonous exhaust gases and made the use of the system impossible around friendly forces. This was solved by using a closed system which enabled it to collect the waste gases in a tank. The system was tested until it was omitted from the SDI program in the 1983-1984 budget. The development of the laser proceeded under the name of the Multi-Purpose Chemical Laser (MPCL), and continued under the U.S. Army with funding from Bell Aerospace Textron.[40]

The development of the Mid-Infrared Advanced Chemical Laser (MIRACL) combined with the Sea Lite Beam Director (SLBD) produced an output of 2.2MW at 3,800 nanometers.[41] The Sea Lite, renamed Sky Lite was the beam steering device for the laser. On 18 September 1987 the laser destroyed several vital components of a Northrop BQM-74 airborne target drone, and caused it to subsequently crash.[42] The drone was flying at a speed of 500 knots and at an altitude of 1,500 feet. In 1989, a Vandal supersonic missile simulating a sea-launched cruise missile was intercepted at a low altitude and a range which was regarded as a real tactical scenario.[43]

In the late 1970s the TRW company built a hydrogen fluorine laser at its San Juan Capistrano facility based on a linear bank of nozzles that yielded 2.2 million watts of continuous wave laser power.[44] The laser was tested in 1978 to destroy antitank missiles. The TRW company again built a chemical laser with nozzles in a cylindrical array under the Alpha project. The original design was fitted with a cylinder 5m by 1m, and yielded around 5MW. In 1980-1981 it was fitted with improved nozzles and yielded 10MW.[45] The Alpha program was designed to demonstrate the feasibility of extrapolating the technology from a 2-3MW chemical laser to 5-10MW output.

In order to target the laser weapons as mentioned previously a tracking device is required. A pointer-tracker was designed between 1979-1983 under the U.S. Defense Department's Talon Gold program. The Talon Gold program was designed to demonstrate advanced acquisition, tracking, and precision pointing and to track targets up to 1500 kilometers with an accuracy of 0.2 microradians, scalable to 0.1 microradians.[46] The device consisted of two telescopes, the first focused the image from the exhausts of the missiles onto an array of electro-optical detectors. The telescope would then move so that the image was at the center of its crosshairs. The control of the pointer-tracker would then progress to a second telescope. This would shine a shortwave, accurate laser beam onto the missile, the reflection of which would be received by its electro-optical detectors, and move to enable it to keep the reflection locked on to the center of its crosshairs. In the laboratory tests the device displayed a capability to point and follow with an accuracy of 0.05 microradians in 1981.[47] Talon Gold achieved the performance levels required for the space-based laser.

In 1991, the Relay Mirror Experiment in space relayed a low-power laser beam from the ground to low-earth orbit and back down to a scoring target board at another location with greater accuracy than required by the space-based laser.[48] The successes in the field of acquisition, tracking and pointing have seen advances in inertial reference, vibration isolation, and rapid retargeting/precision pointing. In 1995 the Space Pointing Integrated Controls Experiment produced near-weapons level results during testing.[49]

The focusing of the laser beam is done by a large mirror. In 1980, the Defense Department's Large Optics Demonstration Experiment (LODE) program contracted to Kodak and Perkin Elmer developed the technology to build large laser-quality mirrors out of segments. This made the transportation of large mirrors possible. The mirror is segmented to enable it to be folded inside a launch vehicle and then unfurled in orbit. The larger the mirror the less powerful a laser that will be required to generate the beam, and hence the less fuel that will be required onboard the satellite. The LODE program was designed to demonstrate a four-meter diameter primary mirror with its associated beam control. United Technologies followed the same assembly method by using segments of composite materials onto which gaseous silicon had been deposited. Lockheed Martin developed minute computer-controlled devices for the rear of the mirrors. These devices ensured that the edges of several of the segments are in alignment. They also enable corrections to be made to allow for any imperfections in the surface of the mirror.[50] The retargeting is achieved by moving the whole of the mirror, since the beam would be distorted and loss of focus would occur if only the individual components were moved. LODE was completed in 1987 and provided the means to control the beams of high-powered lasers.

The Development of the Space-based Laser

The Defense Advanced Research Projects Agency initiated the space-based laser program in 1977, which was subsequently transferred to the Strategic Defense Initiative Organization in 1984, which later became the Ballistic Missile Defense Organization in 1993. In May 1997, a Memorandum of Agreement was executed transferring the space-based laser demonstration program from BMDO to the U.S. Air Force. It is anticipated that improvements in directed energy systems will occur from the development of new approaches to energy transfer, efficient mixing of chemical reactants, scaling of present system to higher power design of wave forms, and more efficient propagation of beams through the atmosphere.[51]

The Alpha HF laser was fired for the first time on 7 April 1989. The test lasted for one-fifth of a second and was part of the Zenith Star space-based laser experiment.[52] The test did not take the laser to its full power, which was estimated at two megawatts, which would have required several seconds to achieve. However, the test verified that the subsystems worked properly and demonstrated that the device and its optics were not damaged by the beam.

The Republican takeover of Congress in 1996 saw a resumption after a two-year hold on the testing of the high-energy laser. The Republican-controlled Congress added $70 million to the $30 million BMDO had requested for space-based laser activities in 1997.[53] This allowed the Alpha/LAMP Integration (ALI) program to proceed without delay. These subsystems consist of a projection telescope with a four-meter aperture known as the Large Advanced Mirror Program (LAMP) and the LODE beam control system. The LAMP verified the mirror in 1989 and it achieved the required surface optical figure, and the mirror was controlled to the required tolerances by adaptive optic adjustments.[54] It consists of a 17mm thick facesheet bonded to fine figure actuators that are mounted on a graphite epoxy-supported reaction structure. The ALI test was conducted at Capistrano and was undertaken by connecting a vacuum chamber containing Alpha to a chamber containing the LODE and LAMP with a seventy-foot beam tube.[55]

A space-based laser was to be flown in a half-scale demonstrator, known as Star Lite as early as 2005. The demonstration test would cost around $1.5 billion.[56] The program Star Lite was born out of the Strategic Defense Initiative's Zenith Star. Prior to Zenith Star's demise in 1993 due to funding and technical problems, it was going to be a 45,000 kg spacecraft with a primary mirror eight meters in diameter. The program was reactivated in 1995 due to breakthroughs in high-reflectivity coatings and adaptive, uncooled glass optics. The new optics reduced Zenith Star to Star Lite. Zenith Star used heavy molybdenum mirrors with active cooling to keep the laser beam from melting the mirrors. The breakthroughs in coatings decreased the amount of energy that was absorbed to a fraction of a percent of megawatt.[57] The Star Lite project was developed by three companies: TRW's Alpha hydrogen-fluorine laser; Hughes Danbury Optical Systems" LAMP; and Lockheed Martin's beam control system.[58] Star Lite was to consist of the optics, the beam control system, the laser engine, and the spacecraft bus which would control the electrics, and would weigh 17,500 kilograms.

In March 1997, TRW and Lockheed Martin completed the first integrated ground test with a 0.5 second long firing of the laser. In July 1997 BMDO declared that the system had successfully sampled its own beam to compensate for jitter, which occurs due to the vibrations of the spacecraft. Star Lite will be limited to around 30-60 seconds of total laser time, using laser fuelled at approximately 30kg/s.[59] Initially it will be tested in 1-5 second bursts in order to check its operation and fired at diagnostic targets to measure how much energy it delivered and discover if the laser maintained its target. Factoring in the funding the program receives and the level of risk accepted in scaling the laser up, it is estimated that an operational space-based laser could be ready about seven years after a successful Star Lite testing.

In March 1998, Boeing and TRW working together as Team SBL (Space-based Laser), were awarded a six-month contract worth $10 million to define the concepts for a Space-based Laser Readiness Demonstrator (SBLRD). The SBLRD was designed to prove the technical feasibility of using a space-based laser to intercept and destroy ballistic missiles in their boost phase. The Team SBL under the contract would define concepts for several issues of the SBLRD program: a concept for the demonstrator space vehicle; a concept for a SBLRD test program; and a risk-mitigation concept.[60] The contract as it was then addresses a fast and normal schedule. The fast schedule envisioned a 2005-2006 launch using existing technologies, whereas the slow schedule had a 2008 launch date and examined newer technologies. Several technologies have been demonstrated which reduce the weight of the space-based laser by 10 percent. These included more efficient rocket nozzles that produce the HF lasing fuel and reduce fuel consumption and lighter weighing spacecraft buses, due to composite materials and better structural analysis.[61]

In February 1999, a contract was awarded by the United States Air Force to a consortium for the space-based laser Integrated Flight Experiment (IFX), previously named the SBLRD. The IFX project was jointly funded by the U.S. Air Force and the Ballistic Missile Defense

Organization. The California-based consortium consists of Lockheed Martin Missiles and Space in Sunnyvale, Boeing in Canoga Park and the Space and Electronics Group of TRW in El Segundo. This represents the first increment of a total contract estimated to be between $2 billion and $3 billion when completed.[62] The first increment of the contract represents $125 million and starts tasks to be undertaken in the first 18-24 months. The Space and Missile Systems Center in Los Angeles run by the U.S. Air Force runs the space-based laser IFX program for the Ballistic Missile Defense Organization (BMDO). The program is designed to assess and progress with the feasibility of the space-based laser concept and its technologies, along with an eventual demonstration in space. It will include ground, flight, and space experiments to verify the technologies at the component and subsystem level. The vehicle provides a center for resolving the challenges from the integration of a system that combines precision optics and high-energy lasers into a lightweight space vehicle.[63] A database will be compiled in order to analyze on-orbit performance, from ground testing of the subscale vehicle.

In total in 1999, $168 million was allocated for the development of the space-based laser.[64] A constellation of twenty laser-firing satellites which would be in orbit 800 miles above the Earth, is envisaged by the U.S. Air Force. A constellation of twenty satellites is the ideal deployment scheme. Other schemes are being considered. One concept would deploy ten satellites and an equal number of orbiting mirrors which would bounce laser beams to their targets. Another alternative is to combine ground-based lasers with an array of orbiting mirrors. One problem with these alternatives is that they require mirrors with diameters of 98 feet, and at present are expensive to construct and extremely difficult to place in orbit.[65] A further technical problem at present is controlling the space-based laser once it is in orbit. At present it is not possible to remotely control the generation of megawatts of power out of a laser.

A twenty satellite constellation at a 40 degree inclination is estimated as being able to provide a robust theater-missile defense capability.[66] At this deployment the kill times per missile will be in the range of one to ten seconds depending on the range of the missile. The retargeting time for new targets requiring small angle changes is 0.5 seconds. A constellation of twelve satellites is able to intercept 94 percent of all missile threats in most theater threat scenarios.[67] A system consisting of twenty satellites would, therefore, provide almost full threat negation. A constellation of twenty satellites operating at an inclination of sixty degree could provide national missile defense threat negation along with full theater missile defense negation.[68] The space-based laser satellite will be equipped with multiple sensors: passive missile detection sensors to locate launching missiles and an active laser-radar to track the missile in its boost phase.

In fiscal year 2000 the Pentagon spent about $237 million on laser technology.[69] In March 2000, TRW operated a test of the Alpha energy laser which produced a 25 percent increase in the laser's power output and improved its quality.[70] The 25 percent power increase was acquired by moving a fixture that prevents stray light from damaging critical electronics inside Alpha out of the laser beam path.[71] The six-second test of the laser was part of the Alpha Laser Optimization (ALO) program, which is funded by the U.S. Air Force and BMDO. The SBL-IFX laser like the Alpha laser will be a cylindrical HF chemical laser, but will benefit from the advances in engineering since Alpha was designed in the mid-1980s. The test increased the amount of power extracted from the cavity of the Alpha laser by the optical systems that focused the energy into a laser beam.[72] The test was part of an approach to explore ways in which the weight and size of the SBL-IFX subsystems can be reduced. The improvement in laser quality and output produced per unit of fuel would allow the size and weight of the chemical fuel tanks to be reduced without foregoing any capability.

A report by the Pentagon published in March 2000 entitled "Laser Master Plan" was concerned by the size of the supplier base which serves high-energy laser technologies. In most sectors of laser weapons the number of suppliers has reduced.

See the table below the U.S. industrial supplier base for key high-energy laser components.

U.S. Industrial Supplier Base for Key High-energy (HEL) Laser Components

Major HEL Components	Prior Suppliers	Current Suppliers	Vendors
Laser Device	6	2	TRW, Boeing
Wavefront Sensors	3	1	Adaptive Optics Associates
Deformable Mirrors	4	1	Xinetics
Coatings	3	2	Barr, Optical Coatings Inc.
Large Mirror Blanks	3	2	Corning, Schott
High Power Windows	2	1	Heraeus

| Focal Plane Arrays (short-wave infrared) | 3 | 2 | Raytheon, Hughes |

Source: U.S. Defense Department and Air Force Research Laboratory

In December 2000 an integrated ground test of the Alpha high-energy laser, its beam director telescope and the beam alignment and correction system provided new information on the monitoring and maintenance of the pointing of the SBL-IFX beam director on orbit.[73] Team SBL conducted a six-second lasing test on 8 December 2000 at TRW's Capistrano test site, as part of the $240 million SBL-IFX development program. The test involved generating a megawatt laser beam with the TRW-built Alpha, which was fed through the beam control system, built by Lockheed Martin, and a four-meter diameter beam director telescope which was housed in a vacuum chamber that simulated the environment of space. The primary goal of the test was to determine whether the telescope's metrology systems could maintain the pointing and proper alignment of the primary and secondary optics during lasing. The test was deemed a success. A secondary achievement of the test was the successful determination of the laser's characteristics such as power, beam uniformity, and frequency spectrum when the laser beam and optical systems were focused on a target.[74]

In the first quarter of 2001, the SBL-IFX program's system requirement review was completed, a forward step in the ongoing design and manufacturing process.[75]

The Team Space-based Laser-integrated Flight Experiment (SBL-IFX) and the Companies Involved

Company	Areas of Responsibility
Lockheed Martin Missiles and Space Operations	Leading development of the SBL-IFX spacecraft and its payload integration; developing and maturing beam director technologies; leading development of the SBL-IFX ground support segment; and leading operational SBL architectural definition
TRW Space and Electronics Group	Leading definition and development of operational SBL-IFX technologies; leading integration of the SBL-IFX payload, developing, and maturing laser payload techniques; and leading development of the test facility

Boeing Space and Communications Group	Leading SBL-IFX systems engineering, integration, and test; developing and maturing SBL-IFX beam control technologies, including those needed for acquisition, tracking and pointing leading optical integration in the SBL-IFX payload segment; and leading the SBL-IFX mission operation segment

Source: Adapted from R. Wilson, "Putting Space Weapons on the Fast Track," *Aerospace America*, July 2001.

The U.S. defense budget's request for 2002 is $110 million for the space-based ballistic missile defense program, that is, the space-based laser program. This space-based laser program may be ready for an initial test in around three years.[76] The chemical hydrogen-fluorine (HF) laser was chosen for the space-based ballistic missile defense role because the HF laser's reactants absorb the waste heat as the reactants are used, and emits the heat into space. The stability and long shelf lives of hydrogen and fluorine are positive factors in their functioning in the space-based laser.[77] The U.S. Air Force anticipate deploying the space-based laser system in around a decade. The major participants in the $4 billion space-based laser demonstration program are Lockheed Martin, Boeing, and TRW.

The experimental demonstration vehicle will weigh between 40,000 and 42,000 pounds and will carry a megawatt-class laser and a 2.8-meter, beam-directing optical mirror. The actual operational system would be equipped with a multimegawatt laser and carry an eight-to-twelve meter mirror. The demonstration vehicle is planned to be in orbit in 2012.[78] The operational space-based laser is to be capable of intercepting ballistic missiles in the upper reaches of the stratosphere (40,000 to 50,000 feet above the earth) and in space. The space-based laser is intended to operate at an altitude of 1,300 kilometers and would have a lethal range of 4,000-5,000 kilometers. A single satellite could cover as much as 10 percent of the Earth's surface.[79] The inability of the laser to penetrate beneath the Earth's atmosphere since the HF laser's effects are diminished by water vapor in the Earth's atmosphere, is considered to be an advantage politically. It consequently does not have the stigma of being a "death-ray weapon" from space to the ground.

The baseline threat against which the space-based laser is being designed is the Russian SS-18 intercontinental missile.[80] The dimensions of the space-based laser IFX are being limited to 53 feet in height and 43,400 pounds in order to fit into the constraints of a heavy-lift Evolvable Expendable Launch Vehicle. The beam control is being assigned

5,681 pounds, while the beam director is accorded 3,420 pounds. In the near term, engineers are pursuing two major paths towards risk reduction. One is the ability to control the laser's wavefront. The manipulation of the wavefront is required during transmission through air (but not in space), to achieve defraction-limited performance, which would allow the system to project sufficient power onto a converged point on the target.[81] The second path is the laser. The Alpha laser does not meet the efficiency requirements and power level demanded for a space-based laser. It is anticipated that a smaller scale laser, known as Short Stack, consisting of 10 of 92 rings which produce the laser energy, will be built in 2003, and will generate significantly more laser time than Alpha.[82]

The construction and design of the Space-based laser Test Facility (STF) is anticipated to start in 2002. The STF is going to be a large facility that will enclose a space-based laser vehicle and enable testing of the entire system, incorporating the high-energy laser in an evacuated space environment.[83] The STF is anticipated to be completed by 2007. The integration of the laser, beam control, beam director, and spacecraft systems would begin in 2006, which would lead to integrated tests in 2008 or 2009. Following on from this a four-year test period in the STF would commence, leading to the launch of IFX from Cape Canaveral in 2012.

The IFX plan has however been subject to budgetary considerations. The decision was taken by the Missile Defense Agency to defund the IFX plan. This has meant that there is no longer a scheduled plan to develop a space-based laser. Instead the space-based laser will be funded on a capabilities-based approach, which means that whichever missile defense system is technically more mature will be given priority funding. This action will nevertheless mean a delay in developing a space-based laser and the probable timeframe for an in-orbit space-based laser is now around 2020. Nevertheless, a space-based laser is still under development in the United States thinking for weaponizing space. The space-based laser continues to be a politically controversial option for ballistic missile defense.

Notes

1. Kosta Tsipis, "Laser Weapons," *Scientific American*, December 1981, 36-37.
2. M. Callaham and K. Tsipis, *High Energy Laser Weapons: A Technical Assessment* (Cambridge: Massachusetts Institute of Technology, 1980), 9.
3. Callaham, 92.
4. Callaham, 17.
5. Callaham, 18.

6. Dietrich Schroeer, "Directed-Energy Weapons and Strategic Defense: A Primer," *Adelphi Papers 221* (London: IISS, Summer 1987), 14.
7. M. A. Dornheim, "Missile Destroyed in First SDI Test at High-Energy Laser Facility," *Aviation Week and Space Technology*, 23 September 1985, 17-19.
8. Ashton B. Carter, *Directed Energy Missile Defense in Space*, Background Paper (Washington, D.C.: U.S. Congress, Office of Technology Assessment, OTA-BP-ISC-26, April 1984), 18.
9. Carter, 18.
10. Carter, 17.
11. Carter, 17.
12. Bengt Anderberg and Myron L. Wolbarsht, *Laser Weapons: The Dawn of a New Military Age*, (New York: Plenum Press, 1992), 91.
13. Bengt, 92.
14. M. Callaham and K. Tsipis, 33.
15. Callaham, 33.
16. Kostas Tsipis, 39.
17. Tsipis, 18.
18. Bengt Anderberg and Myron L. Wolbarsht, 101.
19. M. Callaham and K. Tsipis, 25.
20. Callagham, 25.
21. Callagham, 102.
22. Kosta Tsipis, 39.
23. Ashton B. Carter, 19.
24. Dietrich Schroeer, 27.
25. Ashton B. Carter, 39.
26. Carter, 39.
27. Carter, 40.
28. Keith B. Payne, *Laser Weapons in Space: Policy and Doctrine* (Boulder, CO: Westview Press, 1983), 24.
29. Payne, 24.
30. Payne, 24.
31. Dietrich Schroeer, "Directed-Energy Weapons and Strategic Defense: A Primer," *Adelphi Papers 221* (London: IISS, Summer 1987), 62.
32. Angelo Codevilla, *While Others Build: The Common Sense Approach to the Strategic Defense Initiative* (New York: Free Press, 1988), 143.
33. Uri Ra'anan and Robert L. Pfaltzgraff, *The Internation Security Dimensions of Space* (Hamden, CT: Archon Books, 1984), 66.
34. Angelo Codevilla, 147.
35. Bengt Anderberg and Myron L. Wolbarsht, 108.
36. Bengt Anderberg and Myron L. Wolbarsht, 108.
37. Bengt Anderberg and Myron L. Wolbarsht, 122.
38. Anderberg, 122.
39. Anderberg, 123.
40. Anderberg, 123.
41. Anderberg 123.
42. Anderberg, 123.
43. Anderberg, 124.
44. Angelo Codevilla, 144.
45. Codevilla, 144.
46. Keith B. Payne, 29.

47. Angelo Codevilla, 144.
48. "Space-based Laser (SBL)," *Federation of American Scientists*, http://www.fas.org/spp/starwars/program/sbl.html, accessed October 17, 2001.
49. *Federation of American Scientists.*
50. Angelo Codevilla, 145.
51. "Directed and Kinetic Energy Systems Technology," *DTIC Report*, March 1999, Military Critical Technologies, http://www.iac.dtic.mil/mctl/data/sec04.pdf, 1.
52. "Alpha Missile Defense Laser Is Fired for First Time," *Aviation Week and Space Technology*, April 19, 1989, 23.
53. Joseph C. Anselmo, "New Funding Spurs Laser Efforts," *Aviation Week and Space Technology*, October 14, 1996, 67.
54. Space-based Laser (SBL)," *Federation of American Scientists*, http://www.fas.org/spp/starwars/program/sbl.html, accessed October 17, 2001.
55. Michael A. Dornheim, "Pentagon Mulls Space Laser Test," *Aviation Week and Space Technology*, March 23, 1998, 32.
56. Dave Dooling, 'space Sentries," *IEEE Spectrum*, September 1997, 58.
57. Dooling, 59.
58. Dooling, 58.
59. Dooling, 59.
60. "TRW Wins Space Laser Contract," *Space Daily*, March 19, 1998, http://www.spacedaily.com/nes/laser-98a.html.
61. Michael A. Dornheim, "Pentagon Mulls Space Laser Test," *Aviation Week and Space Technology*, March 23, 1998, 32.
62. "Plasmadynamics and Lasers," *Aerospace America*, December 1999, 15.
63. *Aerospace America*, 15.
64. Frank Vizard, "Return to Star Wars," *Popular Science*, April 1999, 2.
65. Vizard, 2.
66. 'space-Based Laser," *BMDO Fact Sheet*, 301-00-11, November 2000, 2.
67. *BMDO Fact Sheet*, 2
68. *BMDO Fact Sheet*, 2.
69. Robert Wall, "U.S. Laser Weapons Industry Is Shrinking," *Aviation Week and Space Technology*, April 10, 2000, 31.
70. "Megawatt Laser Test Brings Space-based Lasers One Step Closer," *Space Daily*, April 26, 2000, http://www.spacedaily.com.news/laser-00e.html.
71. "TRW Conducts Tests to Validate Laser Technology," *Aviation Week and Space Technology*, May 1, 2000, 29.
72. "TRW Conducts Tests to Validate Laser Technology," 29.
73. "Space-Based Laser Team Advances Design with Successful Test," *Space Daily*, January 25, 2001, http://www.spacedaily.com/news/laser-01a.html.
74. *Aviation Week and Space Technology.*
75. J. R. Wilson, "Putting Space Weapons on the Fast Track," *Aerospace America*, July 2001, 47.
76. John G. Roos, "Militarizing Space," *Armed Forces Journal International*, September 2001, 32.
77. Roos, 35.
78. Roos, 35.
79. Joseph C. Anselmo, "New Funding Spurs Laser Efforts," *Aviation Week and Space Technology*, October 14, 1996, 67.
80. Robert Wall, "Killing Missiles at the Speed of Light," *Aviation Week and Space Technology*, August 14, 2001, 55.

81. Wall, 55.
82. Wall, 55.
83. "Space-based Laser (SBL)," *Director, Operational Test and Evaluation*, United States FY00 Annual Report, http://www.dote.osd.mil/report.FY00/other/00sbl.html, 2.

Chapter Seven

The Revolution in Military Affairs and the Militarization of Space

This chapter examines the Revolution in Military Affairs (RMA) and the militarization of space. The first section examines the impact military space systems made during the Gulf War. The impact this made had profound implications for those claiming an RMA was underway in the United States, Russia, and China. The following section examines the United States approach to the RMA. It examines the key players in promoting the RMA. The elements of the RMA, precision strike, information warfare, and dominant maneuver are analyzed in turn. The underpinning of space systems for the RMA is addressed. The United States section concludes with an analysis of the factors driving the RMA and the concern in the United States of other countries embracing the RMA.

The Chinese approach to the RMA is examined. The individuals involved in the discourse are identified and the origins of the RMA debate in China are traced. Also addressed is the attempt to Sinify the RMA. The Russian approach to the RMA is traced, and indeed is the source of the concept that was identified as the Military Technical Revolution and became the RMA.

The Role of Military Space during the Gulf War

The use of space systems during the conflict in the Gulf to expel Iraq from Kuwait was one of the first occasions that demonstrated the importance that space systems contributed to the air and ground campaign. This section reviews the literature on space systems and the Gulf campaign, and highlights the roles satellites played in adding to the effectiveness of the overall campaign. It is also important to comprehend the political impact the Gulf War had in terms of fueling the debate on the Revolution in Military Affairs, which although a broader concept than the military use of space, benefited significantly from the effectiveness of space systems during the campaign. It is for this reason that the military space literature of the Gulf campaign is assessed.

The war in the Persian Gulf was the first circumstance in which a wide range of military space systems were used in a conflict. It was the first real test under war conditions of the sixty or so Western military satellites that were involved.[1] Space added a fourth dimension to the war. It allowed a communications network to support a 400,000-strong army to be established in theater in a few weeks. It provided images of Iraqi forces and the reconnaissance photographs for the Allied air attacks. Satellites provided a navigation system which provided accurate information for combat soldiers, on missiles, tanks, aircraft, and ships. It is for these principal reasons that the Gulf War is being described as the first space war.[2]

The information provided by satellites offered distinct advantages to commanders planning and conducting ground operations.[3] The commanders on the ground were able to know the enemy force disposition, strength, and the environment where the combat occurred. The theater commander was able to maximize force effectiveness during the battle by having access to information, which gave him knowledge of any changes within his area of operation. Satellites were able to support battlefield preparation, enemy force assessment, targeting, weapons cueing, and battle damage assessment.[4]

Weather monitoring capabilities were able to reduce the "fog of war." During the campaign, the Department of Defense along with commercial meteorological satellite systems were used to acquire reliable weather data over Iraq. The specific kinds of weather information were particularly important for the planning and employment of laser and infrared precision-guided munitions.[5]

> This information was used to determine how best to configure in-theater reconnaissance assets, which precision-guided munitions to employ and when and where a unique-capability force should strike . . . The critical data provided by space-based meteorological systems makes our advanced weapon systems more effective and gives commanders the freedom to exploit the weather as a component of decisive action.[6]

The Global Positioning System (GPS) enabled U.S. forces to navigate using all-weather and day-night accurate positioning information. This allowed the coalition field commander to freely traverse the featureless desert, while the Iraqi forces were limited to roads. The GPS system also supported other functions as well: minefield clearance, artillery fire support, precision-guided munitions employment, and covert missions.[7] JSTARS was used to attempt to track mobile Scud launchers albeit with little success.[8]

GPS handheld receivers were used by ground soldiers to fix their positions. Using a laser range finder the soldier was able to obtain the range and bearing of the target for relay to an air control officer to provide precise target information for ground support aircraft. In fact, GPS receivers it is argued,

> are now credited as making possibly the single most important contribution to the success of the conflict. They certainly saved many Coalition lives and casualties, and significantly enhanced the effectiveness of most teeth arms.[9]

Mapping and charting was a necessary requirement before the onset of Desert Storm, since many of the maps of Iraq and Kuwait were over thirty years old. The earth resource satellites were thus able to provide the information necessary to develop and update the maps. The satellite images enabled simulators to train pilots using images of the actual targets they were going to bomb. In addition to this capability, these resources allowed theater commanders to plan amphibious and airborne operations, track the movements of Iraqi forces, and prepare for and practice strike operations.[10]

Communications satellites carried the majority of the military trunk traffic, such as speech, data, facsimile, telegraph, into and out of the theater.[11] They provided tactical links within theater and bridges for other terrestrial VHF/UHF radio systems whose line-of-site limitations prevented them from spanning the desert reaches. They provided communications to ships at sea, to troops on the move, and even to military aircraft. And when military satellites became overloaded, civil space circuits were leased and pressed into service.[12] Coalition communication systems carried more than 700,000 telephone calls and 152,000 messages per day during the most intense part of Desert Storm.[13] Out of this, satellite communications systems carried 85 percent of the total inter- and intratheater load.[14] At the operational level these systems allowed the coordination of the air, land, space, and special operations forces to be integrated into a comprehensive plan. The rapid transfer of battlefield information from tactical to operational commanders, and then to the strategic level decision makers, was made possible through space-borne communications.[15]

The coalition also had access to satellites that could give early warning of ballistic missiles.[16] These satellites quickly became part of the campaign to detect and shoot down incoming Scud missiles launched against targets in Israel and Saudi Arabia. This Defense Support Program (DSP), was designed to provide early warning of Soviet intercontinental missiles, and was apparently successful earlier in monitoring Iranian Scud launches against Iraq in 1986.[17] The DSP satellites used a twelve-foot telescope to collect infrared energy in an optical sensor, focusing onto an array of two thousand detectors which covered an area of around two square miles.[18] The heat plume of the missile was relayed via satellite and ground stations to the North American Aerospace Defense Command (NORAD), southwest of Colorado Springs.[19]

The detailed workup of the U.S., UK, and French operational space and ground segments in support of the combat forces highlighted the inherent capabilities of space communications. These include rapid deployment, high quality, security, reliability, power, and flexibility. It also demonstrated the growing military dependence on satellite communications in all services and at every level of command. Nevertheless, as Rear Admiral Sir Peter Anson commented,

> Considerable ingenuity was needed to reconfigure the space segment and to manufacture, extend, patch, and adapt and modify ground systems to bring them up to operational scratch over the five months of grace afforded by Desert Shield.[20]

The Iraqis on the other hand had no military space assets.

They did have access to civil international networks, Intelsat and Inmarsat, plus a share in Arabsat, which operates two regional telecommunications satellites covering the area. However, the Arabsat earth station in Baghdad was an early victim of the bombing campaign.

The ability of the Iraqi leadership to have access to the benefits of space imagery was prevented by negotiations with "SPOT Image," the corporation that operates the SPOT series of imagery satellites. The French government agreed not to sell Gulf images to countries outside of the coalition. The space-imagery embargo against Iraq was possible because the few countries capable of providing space imagery of the Middle East region consented to the general aims of the coalition. As Lambakis points out,

> An alternative scenario, however must be considered. Had Soviet leaders believed themselves compelled to provide imagery to their erstwhile ally, the U.S.-led coalition would have had to accept this potentially significant encumbrance or otherwise work to sever the communications links between Baghdad and Moscow or intercept imagery deliveries to Baghdad.[21]

However, a U.S.-based EOSAT, a Spot image competitor that operates the Landsat series of satellites, continued to sell imagery information to concerns in noncoalition countries. It argued that it had a legal obligation to do so.²² Another aspect in this regard is the fact that the United States and its allies used INTELSAT satellites extensively, along with Iraq. There was no attempt made to deny Iraq access to INTELSAT by invoking the INTELSAT charter, because the restrictions could have been applied equally to the United States and its allies.²³ This was due to the fact that if the United States had attempted to deny INTELSAT access to an adversary who was a consortium member, as Iraq was, this could backfire and lead to restrictions on U.S. military use of the system.²⁴

However, the coalition failed in its bid to cut President Hussein off from satellite information. The weak link was the West's appetite for war news. Reporters tested the limits of the U.S. guidelines on coverage of the war. Consequently, Washington reasoned that Iraq would continuously monitor news broadcasts and glean this to military advantage. Lambakis points out that Cable Network News "no doubt became the best source of intelligence for Iraqi leaders."²⁵

The fact that Iraq had no antisatellite systems, enabled the contribution that satellites made to assist the coalition forces to become extraordinarily simple. As Lambakis argues, "as a consequence of the growing reliance on satellites to perform military functions, a future ASAT-wielding adversary of the United States might be capable of leveraging a victory out of otherwise hopeless military circumstances."²⁶ The integration of space systems within the U.S. military operations may lead to space rapidly becoming the country's Achilles" heel.²⁷

The Gulf War demonstrated that space is increasingly becoming an important arena in war. If Iraq had possessed the ability to use the GPS satellites, its ballistic missile tactics might have been far more devastating.²⁸ The possibility that in the future the United States may encounter an enemy with an ASAT capability has to be considered in the planning for future conflicts. The use of space in providing information for C4I means that space could consequently become another dimension that can affect the outcome of war.

The Revolution in Military Affairs and the United States

The Gulf War and the capabilities it demonstrated saw the development of a new concept, one which was termed the "Revolution in Military Affairs." This term was seized upon by Andrew Marshall, the head of Net Assessment in the Pentagon, who had been sponsoring studies in this area for some time.²⁹ Indeed, at the time the U.S. Secretary of Defense, Richard Cheney in regard to the RMA and its potential claimed that this had been demonstrated dramatically.[30] However this view is called into question by Stephen Biddle in a recent article:

> Rather than a revolution through information dominance and precision strike, what the Gulf War really suggests is a new ability to exploit mistakes. This, however, suggests, very different policies. If new technology offered tremendous military power to any who acquired the new systems (and reformed their military doctrine to exploit them), this implies a powerful incentive for radical change: those who realize the full potential of the new era would enjoy enhanced security and influence, while those who do not do so risk being left behind.[31]

The revolution in military affairs is based primarily on the impact made by the advancement of technologies in the field of information technology, sensors, computing and telecommunications, and the modern military. The concept is defined in the Annual Report to Congress as:

> A Revolution in Military Affairs (RMA) occurs when a nation's military seizes an opportunity to transform its strategy, military doctrine, training, education, organization, equipment, operations, and tactics to achieve decisive military results in fundamentally new ways.[32]

The interplay of advanced technology and new operational concepts can occur in two distinct ways. The first is the requirements pull, where a new critical operational task emerges which requires the development of new technology to accomplish the new mission. An example of this would be ballistic missile defense, where the proliferation of ballistic missiles and its associated technology created the requirement for theater missile defense for troops and potentially national ballistic missile defense. The second is the technology push, where a promising new technology spurs the development of a new weapons system or operational concept and enables a new mission to be performed. An example of this would be the utilization of the global positioning system to navigate for precision munitions. It is the combination of the requirements pull and technology push that has seen the maturation of technology providing new missions which has contributed to the RMA.

One of the most influential advocates of this approach in the United States Defense establishment was Admiral Owens. Owens' approach is commonly known as the 'system of systems" approach. This approach is based upon system architectures and joint operational concepts. It attempts to gather real-time, all-weather information continuously to formulate dominant battlespace knowledge.[33] Dominant battlespace knowledge (DBK) involves relaying automated target information to operational plans into a recognized network in which the information can be utilized. This infusing of DBK into U.S. forces combined with the real time awareness, it is argued, will enable the United States to apply force with speed, accuracy, and precision.[34] The conflict in Iraq is often cited as a demonstration of these capabilities. As Admiral Owens espoused at a conference:

> If you see a battlefield the size of Iraq, if you start to have the technical capability to provide a tactically fused knowledge base to that battlefield, you can go after the targets that are the strategic center of gravity in the battlefield, and you will win. It means that you do not have to have as many tanks, ships, or airplanes. It means that you can put together a whole new theory of the way you fight wars.[35]

Those skeptical of the RMA point to the Clausewitzian trinity of primordial violence, chance and probability, and subordination of war to policy, and see these concepts being replaced by a new technological trinity: intelligence, surveillance, and reconnaissance technologies; advanced command, control, communications, and computer systems; and precision strike munitions.[36] The RMA concept is being formulated independently from an adversary in a closed environment where friction, chance, and uncertainty are omitted. The problem is not with the technology but lies with the view that a lead in technology by itself is enough, and investment in those emerging technologies could be to the detriment of force structure, readiness, and training. This point is illustrated by the following:

> The insertion of large numbers of troops capable of holding their ground into critical positions remains a vital military task. . . . It is also a task upon which new technologies may have little impact. Sheer numbers and raw military power can be valuable precisely because they are conspicuous.[37]

The Elements of the RMA

The first of the elements of the RMA is precision strike. The concept has its origins in the use of laser-guided bombs in Vietnam, but the guidance systems have since been developed. The technologies have been coupled to the intelligence collection, communications, data processing, and command and control systems as equally necessary components in achieving precision strike.[38] These improvements make precision strike a crucial factor on the modern battlefield.

The second element of the RMA is the concept of information-led warfare. The information technologies make it possible to gather, process, and move enormous amounts of information extremely quickly. This can make it possible for the military commander to formulate a near-real time disposition of the enemy forces and to be able to replan and direct forces accordingly.[39]

The third element of the RMA is dominant maneuver. This concept is a corollary to precision strike and information warfare in that the detailed information on the enemy combined with precision strike can be brought to bear on the enemy's weak points and could result in dominance on the battlefield.[40]

The RMA is underpinned by space systems that provide much of the information which is utilized when acquiring an illumination of the battlefield. Space provides the arena for the sensors, and the transmission through the sensors of the information that allows the battlefield commander to know much more than the opponent. If the ability to utilize these space systems is hampered then the enabling edge these assets provide will be affected detrimentally. Although there are other ways to deploy sensors, space offers superior performance.[41] Indeed, the recognition of the increasing importance of space is made clear in the Annual Report to Congress when it declares. "DOD must be able to ensure freedom of access in space for friendly forces and, when directed, limit or deny an adversary's ability to use the medium for hostile purposes . . . DOD must have the appropriate capabilities to deny when necessary an adversary's use of space systems to support hostile military forces."[42] Implicit within this is that the capability to use space for the benefits it provides is the key element of the RMA.

Factors Driving the RMA

The implementation of the RMA was announced in the publication of the document entitled "Joint Vision 2010." This emphasizes the role of technology in the developing doctrine of the U.S. forces. It follows on from the system of systems approach when it states that, "Technological advances will continue the trend toward improved precision. Global positioning systems, high-energy research, electromagnetic technology, and enhanced stand-off capabilities will provide increased accuracy and a wider range of delivery options."[43] The document attempts to coordinate the advances in computer processing, precise global positioning, and telecommunications to determine the locations of friendly and enemy positions and distribute this data to thousands of locations.

Joint Vision 2010 espouses four areas of continuing superiority which include the aforementioned elements of the RMA concept, but add the concepts of full-dimensional protection (defense against all threats regardless of their source) and focused logistics (the ability to get the right material to the warfighter).[44] Implicit within the concept of full-dimensional protection is the concept of ballistic missile defense, with the ability to defend against theater ballistic missiles, and perhaps national missile defense. This is later made explicit in the Annual Report to Congress:

> Development and deployment of a multi-tiered theater missile defense architecture, combined with offensive capabilities to neutralize enemy systems before and immediately after launch, are prime examples of full-dimensional protection efforts.[45]

One of the greatest motivations behind the RMA concept is that in the face of shrinking defense budgets and hence military force structures, the United States may have few options but to utilize civilian information technologies to increase the effectiveness of their armed forces.[46] In fact this is one of the major differences with the current RMA and previous military revolutions in that the commercial sector is playing a critical role in developing technology which has military applications. This is true with regard to areas such as space launch, navigation, and reconnaissance. The budgetary aspects of commercial space systems and their roles in providing information of the battlefield is illustrated by Admiral Owens:

> It is the unmanned aerial vehicle flying over Bosnia today called the Predator with 2000 hours of proven flight time that provides you with the real-time video through a commercial satellite because it is too expensive to use a military satellite on a hour to hour basis. We prefer to use Motorola satellite because we can rent it cheaper and we can get the downlinks anywhere in the world, and it is much more efficient way to do business with COTS (Commercial Off The Shelf Technologies).[47]

The implications of this means that the military cannot control the spread of technology. Indeed the recent Cox Report has demonstrated the antagonisms between the Commerce Department and the Defense Department over the export of space launch technologies.

The concept of the RMA has become particularly useful politically, especially with the costs associated with maintaining a Cold War-sized military structure. There is the notion that the RMA provides the opportunity to create a new force multiplier that will enable the military to do more with less.[48] Indeed there have been suggestions that the United States should restructure its military and accelerate the integration of the new technology into the active forces.[49] This would see a reduction in the annual defense budget from $245 billion to $210 billion by early in the next century.[50] This does seem rather paradoxical since as Edward Luttwak points out, "it is invariably expensive to reduce casualties by developing and deploying ways to move personnel and materiel farther from harm's way on the battlefield while maintaining their effectiveness."[51] While the issue here is not reducing casualties, the sentiment of the increased cost of moving away from the battlefield and utilizing precision-guided weaponry (one of the key elements of the RMA), does seem to imply an added expenditure, not reductions.

In the post-Cold War environment practical and ethical constraints may impinge more heavily on the United States, focusing upon leaders or peoples, or their means of subsistence. The conduct of the war during the Gulf War is a case in point. The use of information technologies and the new space-based systems which supported them both reduced the number of casualties to the coalition forces, and kept Iraqi civilian losses down.[52] Indeed recently, in the face of Iraqi resistance to weapons inspection there have been calls for a policy of offensive information warfare against Iraq.[53] This may then lead to an option being utilized that has its origins in the present element of the RMA, namely information dominance. This would consist of targeting the military and infrastructure components of the enemy and their links to the other key elements.[54]

An interesting insight into the motivations behind the RMA is provided by Freedman when he argues that "in practice, the revolution in strategic affairs is driven less by the pace of technological change than by uncertainties in political conditions."[55] Indeed since the demise of the Soviet Union there has been a great deal of uncertainty in international security affairs as to what is the greatest issue of concern, and indeed this lack of focus may have contributed to the support for the idea of a RMA. However, the military technological developments, especially the contribution of military space systems do appear to offer and provide new missions which were previously unobtainable. The combination of technology push, for example with GPS for precision munitions as mentioned above, and new missions to deal with the proliferation of ballistic missiles and their associated technologies, do seem to discredit the claim that the RMA is driven more by political uncertainties than technological factors.

The United States Concerns Regarding Other Military Powers and the RMA

One of the issues related to the RMA is the possibility for other military powers to be able to omit a few stages in military development. The most feasible of the elements of the RMA open to a developing military capability is that of precision strike weapons. Although intelligence projections raised the specter of Third World precision-guided munitions twenty years ago,[56] the limited Global Positioning System access coupled with commercial quality reconnaissance data presently available could see this capability finally arrive. Also, the combination of nuclear weapons and some elements of the RMA, such as crude GPS fixes, could reduce the need for extreme position in navigation, target geolocation, and target characterization.[57]

The potential for other military powers to counter the benefits that may accrue from the RMA lies in the vulnerability of the space systems that underpin it. For example, the ability to jam GPS satellites, communication satellites, and particularly commercial satellites, is relatively simple.[58] Developing nations, if their interests are mainly regional, could utilize cheaper technologies such as fiber-optic land lines and direct line-of-sight terrestrial relays for communications for their own military requirements.

It is conceivable that an aspiring global power could radically restructure their military forces in ways that would lead them to acquire forces in a completely different manner to that of the major powers, but it could be costly. For example, a country which has some relatively modern technical capability, such as India, may forego developing a traditional navy and air force and invest in space forces, such as armed transoatmospheric vehicles and perhaps armed satellites.[59] This could give both regional and global strike capabilities greater than it would otherwise achieve with conventional capabilities. However, there are a number of difficulties with this scenario not the least in terms of technical and operational. Also it is more than likely that any aspiring nations would face parochial interests within each of the armed services ready to oppose such proposals.

Russian Views of the Revolution in Military Affairs

The Soviet Union in the late 1980s came up with the term coined by Marshal Nikolai Orgarkov, Military Technological Revolution. Ogarkov believed a fundamental change had occurred in military affairs with the potential of conventional weapons increasing at least tenfold.[60] The United States was seen as at the vanguard of this Military Technological Revolution with its conventional capabilities. The term MTR was supplanted by the use of RMA by Pentagon officials who were familiar with Soviet military theory.

The Russians anticipated between 2000 and 2010 or 2015 the deployment of directed energy weapons, earth-penetrating weapons, and advanced robotics.[61] The technological trend has reprioritized quality over quantity in future military development. The previous qualitative-quantitative argument has been superseded with the decisive impact being dependent on the development of new design concepts and prototypes.

Russian views of future war are expected to be global in their aspirations and they stress that control of space will be the decisive determinant in operations concerned with controlling sections of the earth.[62] The characteristic of war is deemed to have altered. Large quantities of ground troops will no longer be employed. They will be replaced with substantial strikes delivered by remotely piloted precision-guided weapons. A country will be subjugated to precision strikes and will be a battlefield in war without flanks. The distinction of front versus rear will be replaced with that of targets and nontargets, in that there will be no clearly drawn battle lines. Conventional assets will be able to achieve strategic objectives. The Russians declared the Gulf War as the first of the technological operations. The ability of advanced nonnuclear technology to accomplish missions previously earmarked to nuclear forces means that these assets will achieve the objectives envisioned in a nuclear war.[63] These aims will be achieved without the collateral damage and political considerations associated with nuclear weapons.

The Russian military views outer space as a potential theater of military actions. The forms of operations that will be conducted in near-earth space will incorporate the following: operations to destroy strategic weapons in flight; operations to destroy or prevent deployment of enemy satellites; operations to defeat orbital and ground space groupings and to seize and hold strategically important spheres of the near-earth space; and strikes delivered from space.[64]

Chinese Views of the Revolution in Military Affairs

As early as 1988, Chinese writers had shown interest in the revolution in military affairs. Indeed General Mi Zhenyu had discussed the issue in his book that was published in 1988 entitled, "Chinese National Defense Concepts."[65] Interest increased in 1995 with Chinese writers referring to the "third military technical revolution," the Russian terminology as opposed to the U.S. term of RMA. The "Liberation Army Daily," the official newspaper at the time began to publish weekly articles concerning the topic, and in October of that year a national conference was convened to consider the implications for China.[66]

Those who advocate the RMA are represented by a small portion of the PLA, though they include strategists in premier academic institutions, officers in Commission of Science, Technology, and Industry for National Defense, COSTIND, the Second Artillery (the strategic missile unit) and a few others equipped with modern cruise missiles.[67] However, what this group of advocates is missing is a senior Chinese leader to enhance their position. China in May 1996 formed a strategic research center to coalesce research on traditional Chinese statecraft with studies designed to generate innovative military concepts.[68]

Among the RMA advocates, space warfare is seen as being central to the determination of future wars. Chinese thinking regarding space warfare seems very disparate. They appear to amalgamate the following elements: antisatellite weapons, ballistic missile defense, satellite miniaturization, and satellite launcher. However, there does not appear to be any overarching strategy to combine these together with regard to RMA thinking. Tsien Hsue-Shen, the father of the Chinese space program in 1985 brought to the attention of the senior military leadership Russian work on the RMA. Some Chinese strategists envision space weaponsystems as the third weapons revolution, following from conventional and nuclear weapons.[69] The most enthusiastic supporters of the RMA are strategy planners in the headquarters of the PLA's specialized services, and academic staff in PLA research institutions.[70] Their opinions on the RMA are grounded on their interpretation of technological developments in the new century and this in turn has won them support from the Central Military Commission (CMC) chairman, Jiang Zemin. These supporters are at present young, but nevertheless well placed in their career paths to become more influential in the future. Jiang Zemin himself is a farsighted technocrat and is extremely enthusiastic regarding the RMA.[71] Indeed, Jiang Zemin on an inspection tour of the PLA National University of Science in 1991 declared that any future war would be a war involving high technology, a war of multiple dimensions, a war of electronics, and a war of missiles. The PLA would have to be ready for such an occasion.[72] A "qualitative construction of the military" in order to cope with modern local war was formally adopted by an enlarged session of the CMC in December 1991.[73]

Chinese authors on the RMA stress that their country must discover its own "unique techniques and skills" during its examination of the RMA. Hence, it must not simply transfer Western thinking and add Western developments to its existing framework:

> Due to their different economic and scientific development levels, as well as their different cultures, traditions, and ways of thinking, different countries will be subjected to different impacts produced by military revolutions; as a result, they will adopt different approaches toward new things and accept the new military revolution in varying degrees. Therefore there will be a growing trend toward diversification in the pattern of war at the initial stage of the military revolution.[74]

However, at present China has an equipment vacuum in its technological capabilities in order for it to fully transform its forces into an RMA fighting force. At present the RMA school of thought can be seen as a philosophical blueprint rather than a practical indicator for China's defense.[75] When China's national strength grows, so will the RMA's influence on the PLA's military preparations, and it will therefore impact on the high command's thinking on the subject.

In 1998 at a PLA National Defense University workshop, the concept of the RMA was agreed upon, that the RMA is composed of five revolutions: the military thinking of officers, military technology, military equipment, strategic theory, and force structure.[76] They also established that the core of the RMA was the rapid development of information technology. The Chinese are mindful of the need not to merely copy an RMA new force structure and combat patterns of their prospective opponents. To this end the Chinese are attempting to Sinify the RMA concept. They believe that the RMA is in its infancy, with its initial phase extending to 2030.[77] With this in mind it is therefore difficult to conceptualize the RMA's full potential. Though a driving force for the RMA is new technology, this alone without new combat theories will not be decisive in any future conflict. The RMA is seen as providing pressure on the PLA to remove the historical burdens of the revolutionary ideology and outdated military strategies. The RMA has also made China address the practices and strategies of its potential adversaries.[78] By analyzing adversaries strategies, this could allow the PLA to develop its own combat theories.

The RMA has been an incentive for the development of science and technology for China. China's rapid economic growth has enabled it to make research into high technologies more affordable. Combining this desire for technological development, both civilian and military leaders have decided to make a national effort to boost China's defense technology as quickly as possible. An outcome of this is that research and development have focused on developing a military space network, fixed energy and laser equipment, electronic weapons, and super-computers.[79]

A visible expression of China's commitment to the RMA concept is the creation in 1998 of a General Equipment Department (GED) which is under the CMC, with the same rank as the General Staff Department (GSD).[80] This new department has taken over the functions of weapons research and development, testing, acquisition, allocation, and other related matters previously assumed by top agencies in the PLA headquarters. The creation of the GED is significant in that it is an effective measure to create a high-technology defense strategy and without advanced technological weaponry the RMA will not progress. Additionally, the GED is a first step in transforming the theoretical concept of the RMA into actual practical equipment, for instance the department will develop weapons acquisition in line with the requirements of the RMA. The PLA's long term weapons development program sets out three aims.[81] The first aim is the research and development for weapons systems designed for the defense of China's periphery, where low-level conflicts are anticipated. The second is research and development for advanced conventional weapons, for conflicts within or outside of China's territory against high-technology countries. The third, is research and development for strategic high-technology weapons, which include space weapon systems.

The importance of space to China's perception of the RMA cannot be overstated. It is a fundamental aspect of the PLA's long-term research and development in its pursuit of military power. The use of space assets such as navigation satellites to enhance the precision of weapons is a case in point, where the weapon's effectiveness is increased. The following quote highlights this: "A reliable and substantial space capability is now viewed not only as a multiplier of military power, but also as an indispensable factor in organizing united operations and a crucial deterrent to potential adversaries."[82] Indeed this view has been one held by some Chinese strategists from the early 1980s.

The following quote highlights China's focus on high-technology weaponry:

> China's post-Deng strategy is forward-leaning in both political and military terms. Politically, the high-tech focus aims at defense against strategic concerns, namely the major military powers. At the same time the strategy is flexible in principle, catering to different scenarios, from major high-tech wars to small scale border rifts. This is the response of China's armed forces to the country's changing security environment.[83]

China recognizes that there is a gap between its technological capabilities compared with the United States and other NATO countries. In order to narrow this gap, the PLA has focused on a number of high-priority military technological areas. These are seen as fundamental to elevating the PLA's technological level. They are electronic warfare, precision-guided missile technology, high speed computers, powerful laser facilities, and the application of artificial intelligence in military facilities.[84]

One of the central themes emanating from Chinese views of future warfare is the importance of military space. As some authors argue, as space technology develops, the deployment of space-based weapons will make mastery, or control of space a prerequisite for naval victory.[85] These authors envisage space-based weapons, combined with reconnaissance and elint satellites to monitor and track the ships intercepting them. Other authors cite military space's importance, but in relation to air and space. They make the claim "link air and space forces together, under the strategic principle that the one who controls outer space can control the Earth, super powers and military giants are expanding their strength in outer space and the function of air force."[86]

Conclusion

The extensive use of space assets during the campaign to liberate Kuwait demonstrated the importance of space assets to modern war. The array of space systems, such as GPS, communications satellites, and photoreconnaissance satellites enabled the battlefield commanders to illuminate the battlefield on a scale never seen before. These hostilities were the setting for the RMA debate, not only in the United States, but for analysts in China and Russia as well. The performance of the United States in particular in the campaign led Russian and Chinese defense analysts to examine whether there was a novel way of conducting war.

The United States, Russia, and China all identify military space as playing a fundamental part in achieving an RMA. The components that comprise the RMA are precision strike, information warfare, and dominant maneuver. These components are underpinned by space systems that supply much of the information that is required to illuminate the battlefield in order to take advantage of the above components. Space is the arena in which the sensors, and the transmission of the sensors of the information operate. These assets enable the battlefield commander to have the edge over an adversary. It is not altogether surprising to discover that the United States, along with China and Russia have recognized the importance of space systems to the current RMA.

One effect the RMA could have given the central importance of military space systems to the RMA is that it could make space a battlefield. As other countries, such as Russia and China see the effects of the RMA and its devastating effects it has on the battlefield, this may lead them to target the space systems themselves. The reliance on space systems to promote the RMA could lead to space becoming a battlefield.

Notes

1. Sir Peter Anson, Rear Admiral, Royal Navy retired and Dennis Cummings, Captain, RAF. retired, "The First Space War: The Contribution of Satellites to the Gulf War," *RUSI Journal*, Winter 1991, 45.
2. Anson, 45.
3. Dale R. Hamon, Commander, U.S. Navy, retired and Lieutenant Colonel Walter G. Green III, U. S. Air Force, retired, 'space and Power Projection," *Military Review*, November 1994, 63.
4. Hamon, 63.
5. Steven Bruger, Lt. Col., USAF, "Not Ready For the First Space War—What about the Second?" *Naval War College Review*, Winter 1995, 77.
6. Bruger, 64.
7. Anson and Cummings, 48.
8. Michael R.Gordon, and General Bernard R. Trainor, *The Generals' War*, (New York: Little, Brown and Co., 1994), 237.
9. Anson and Cummings, 50.
10. Hamon and Walter, 64.
11. Anson and Cummings, 45.
12. Andson, 45.
13. Steven Bruger, 75
14. Bruger, 76.
15. Bruger, 76.
16. Steven Lambakis, "Space Control in Desert Storm and Beyond," *Orbis*, Summer, 1995, 419.
17. Anson and Cummings 51.
18. Bruger,78.
19. Bruger, 78.
20. Bruger, 46.
21. Lambakis, 421.
22. *The Washington Post*, February 19, 1991.
23. Daniel Gonzales, *The Changing Role of the U.S. Military in Space* (Washington, D.C.: RAND, 1999), 35.
24. Gonzales, 35.
25. Lambakis, 422.
26. Lambakis, 425
27. Lambakis, 425.
28. Lambakis, 426.

29. L. Freedman, "Revolution in Strategic Affairs," *Adelphi Papers*, 318, IISS, 1998, 32.

30. Department of Defense, *Conduct of the Persian Gulf War, Final Report to Congress* (Washington, D.C.: USGPO, 1992), 164.

31. Stephen Biddle, "Victory Misunderstood: What the Gulf War Tells Us about the Future of Conflict," *International Security*, Fall 1997, 176.

32. William S. Cohen, Secretary of Defense, *Annual Report to the President and the Congress*, Department of Defense, United States of America (Washington, D.C.: U.S. GPO, 1999), 122.

33. William A. Owens, Admiral, "The Emerging U. S. System of Systems," National Defense University, *Strategic Forum*, Number 63, Institute for National Strategic Studies, February 1996.

34. Owens.

35. William A. Owens, Admiral, "Revolution in Military Affairs: U.S. Vision for Future Warfare," *Revolution in Military Affairs? Challenges to Governments and Industry in the Information Age*, Conference held at the Royal Institute for International Affairs, 21 and 22 May 1997, 3.

36. Mackubin T. Owens, "Technology, the RMA, and Future War," *Strategic Review*, Spring 1998, 67. See also, Williamson Murray, "Clausewitz Out, Computer In Military Culture and Technological Hubris," *The National Interest*, Summer 1997, 63.

37. Lawrence Freedman, "Britain and the Revolution in Military Affairs," *Revolution in Military Affairs? Challenges to Governments and Industry in the Information Age*, Conference held at the Royal Institute for International Affairs, 21 and 22 May 1997, 5.

38. Glenn C. Buchan, *The Impact of the Revolution in Military Affairs on Developing States' Military Capability* (Santa Monica, CA: RAND, July 1995), 9.

39. Buchan, 9.

40. Buchan, 9.

41. C. S. Gray, "A Contested Vision: The RMA Debate Today," *Revolution in Military Affairs? Challenges to Governments and Industry in the Information Age*, Conference held at the Royal Institute for International Affairs, 21 and 22 May 1997, 6.

42. William S. Cohen, Secretary of Defense, *Annual Report to the President and the Congress*, Department of Defense, United States of America, (Washington, D.C.: U.S. GPO, 1999), 86.

43 *Joint Vision 2010*, Office of the Chairman of the Joint Chief of Staffs, Joint Staff, (Pentagon, Washington, D.C., April 1996), 11.

44. Martin C. Libicki, "Information War, Information Peace," *Journal Of International Affairs*, vol. 51, no. 2, Spring 1998, 416.

45. William S. Cohen, Secretary of Defense, 124.

46. Glenn C. Buchan, *One-and-a-Half Cheers for the Revolution in Military Affairs* (Santa Monica, CA: RAND, 1998), 26.

47. William A. Owens, *Revolution in Military Affairs? Challenges to Governments and Industry in the Information Age*, 4.

48. John Arquilla, "The Strategic Implications of Information Dominance," *Strategic Review*, Summer 1994, 30.

49. See James R. Blaker, "Understanding the Revolution in Military Affairs: A Guide to America's 21st Century Defense," *Progressive Policy Institute*, January 1997.

50. See James R. Blaker, "Understanding the Revolution in Military Affairs: A Guide to America's 21st Century Defense," *Progressive Policy Institute*, January 1997. See also, James R. Blaker, "A Vanguard Force: Accelerating the American Revolution in

Military Affairs," *Progressive Policy Institute*, Defense Working Group Policy Brief, November 1997.

51. Edward N. Luttwak, "A Post-Heroic Military Policy," *Foreign Affairs*, July/August 1996, 40.

52. Lawrence Freedman, *Sanctuary or Combat Zone? Military Space in the 21st Century*, Air Power and Space—Future Perspectives Conference, September 12 and 13 1996, 5.

53. Andrew Rathmell, "Mind Warriors at the Ready," *The World Today*, The Royal Institute of International Affairs, November 1998, 290.

54. John Arquilla, "The Strategic Implications of Information Dominance," *Strategic Review*, Summer 1994, 28.

55. Lawrence Freedman, *The Revolution in Strategic Affairs*, Adelphi Paper 318 (London: International Institute for Strategic Studies, 1998), 76.

56. Glenn C. Buchan, *The Impact of the Revolution in Military Affairs on Developing States' Military Capability*, 12.

57. Buchan, 18.

58. Buchan, 16.

59. Glenn C. Buchan, *One-and-a-Half Cheers for the Revolution in Military Affairs* (Santa Monica, CA: RAND, 1998), 23-24.

60. Robert R. Tomes, "Revolution in Military Affairs—A History," *Military Review*, September/October 2000, 101.

61. Mary C. Fitzgerald, "The Soviet Military and the New 'Technological Operation' in the Gulf," *Naval War College Review*, Autumn, 1991, 18.

62. Fitzgerald, 21.

63. Fitzgerald, 38.

64. Mary C. Fitzgerald, "The Russian Military's Strategy For 'Sixth Generation' Warfare," *Orbis*, Summer 1994, 461.

65. Michael Pillsbury, *China Debates the Future Security Environment*, (Washington, D.C.: National Defense University, 2000), 264.

66. Pillsbury, 264.

67. Pillsbury, 275.

68. Pillsbury, 288.

69. Mel Gurtov and Byong-Moo Hwang, *China's Security The New Roles of the Military* (London: Lynne Rienner, 1998), 91.

70. You Ji, "The Revolution in Military Affairs and the Evolution of China's Strategic Thinking," *Contemporary Southeast Asia*, December 1999, 348.

71. Ji, 348.

72. Ji, 353.

73. Chenug Tai Ming, "Decimated Ranks: Peking Further Cuts PLA in Bid to Modernise Military," *Far Eastern Economic Review*, February 27, 1992, 15.

74. Zhang Feng, "Historical Mission of Soldiers Straddling 21st Century," *Jiefangjun Bao*, January 2, 1996 as translated in FBIS-CHI-96-061, quoted in Mr. Timothy L. Thomas, "Behind the Great Firewall of China: A Look at RMA/IW Theory From 1996-98," *Foreign Military Studies Office*, November 1998, 4.

75. You Ji, "The Revolution in Military Affairs and the Evolution of China's Strategic Thinking," *Contemporary Southeast Asia*, December 1999, 349.

76. Ji, 349.

77. Ji, 351.

78. Ji, 352.

79. Ji, 357.

80. Ji, 359.
81. You Ji, *The Armed Forces of China*, (I. B. Taurus: New York, 1999), 56.
82. Ji, 78.
83. Ji, 8.
84. Ji, 58.
85. Captain Shen Zhongchang, Lt. Com. Zhang Haiying, and Lt. Zhou Xinsheng, "21st Century Naval Warfare," in Michael Pillsbury, ed., *Chinese Views of Future Warfare* (Washington, D.C.: National Defense University, 1998), 263.
86. Maj. Gen. Zheng Shenxia and Senior Col. Zhang Changzhi, "The Military Revolution in Air Power," in Michael Pillsbury, ed., *Chinese Views of Future Warfare*, 308.

Chapter Eight

The Post-Cold War Military Space Policy of the United States

This chapter examines the United States' military space policy since the end of the Cold War. It analyzes President Clinton's two terms of office and the start of President Bush's administration with respect to missile defense policy and military space policy. The Clinton administration's period in office saw political maneuvering between Congress and the president over national missile defense plans. There were a number of congressionally initiated acts to instigate a program towards the building of a national missile defense system. This chapter examines these architectures with regard to the use of an exoatmospheric interceptor, which intercepts the threat ballistic missile in space. The chapter progresses to examine the impact of the Presidential Directives that were announced during the Clinton period, on military space policy. The organization changes that were implemented with regard to military space are analyzed.

The following section examines President Bush and the missile defense policy that he has espoused. It examines the significance and rationale for the United States' withdrawal from the Antiballistic Missile Treaty. The Commission to Assess the U.S. National Security Space Management and Organization reported during the first months of President Bush's administration. The impact this had on military space policy and the organizational changes it had on the space infrastructure are analyzed. In this section the space-based weapons that are being considered are outlined with particular attention given to space-based weapons against terrestrial targets. The chapter finally assesses the impact and the contribution military space assets have made to recent conflicts. It examines the roles space assets made to the campaign in Yugoslavia, and the events in Afghanistan.

President Clinton's Missile Defense Policy

The Clinton administration in 1993 inherited the Global Protection Against Limited Strikes ballistic missile defense system and promptly cancelled the development of the system. The priority was placed on the development of theater missile defense with national missile defense placed into research and development. This reorientation of missile defense policy was reflected in the name change in May 1993 from the Strategic Defense Initiative Organization to Ballistic Missile Defense Organization, indicating the shift from strategic defense to theater missile defense. The refocus was primarily to concentrate on ground-based defenses with a reduced effort towards national missile defense.

The quest for a National Missile Defense of the United States was undertaken by the Republican-controlled Congress. The Congressional attempts to legislate for a national missile defense gave rise to a number of Missile Defense Acts. The first of these was the Missile Defense Act of 1995. This act called for the deployment of a ground-based National Missile Defense with multiple sites to be operational by 2003.[1] The act envisioned the deployment of up to one hundred ground-based interceptors, supported by space-based sensors. Also, contained with the act was U.S. intent to negotiate treaty changes with Russia, and if those negotiations failed, to withdraw from the treaty. However the Missile Defense Act of 1995 was unacceptable to the Clinton administration and a bipartisan compromise was achieved. This watered down the previous provision and committed a ballistic missile defense system to be developed for deployment, and deployed only if Russia approved. If no approval was reached, the option of withdrawal from the treaty would be considered. The Secretary of Defense would develop an interim national missile defense plan that would give the United States the ability to field an operational capability by the end of 1999 if required by the threat. This compromise was approved on September 5, 1995.

The following year the Republican-controlled Congress initiated the Defend America Act of 1996. The Defend America Act declared that it was the policy of the United States to deploy at the end of 2003 a National Missile Defense system that was capable of providing a highly effective defense against limited, unauthorized, or accidental ballistic missile attacks, and would be augmented over time to provide a layered defense against larger and more sophisticated ballistic missile threats as they emerged.[2] Unlike the Missile Defense Act of 1995, this act did not specify or explicitly restrict the United States to ground-based systems; yet it did not explicitly challenge the ABM Treaty. The Congressional Budget Office estimated that the deployment of the proposed system would cost either thirty-one or sixty billion dollars, which led to the Defend America Act being withdrawn.[3] The attempt by the Republicans in the 1996 Presidential election year to make national missile defense a major issue during the campaign did not materialize. However, President Clinton and Congressional Democrats had taken this issue seriously enough to adopt a compromise in the form of a deployment plan called the "3 plus 3" plan.

The National Missile Defense "3 plus 3" Program

In reaction to Congress and in concern about the emergence of a ballistic missile threat to the United States sooner than the Intelligence Community projections, the Ballistic Missile Defense Organization (BMDO) commissioned a Tiger Team study to identify feasible alternatives for a treaty-compliant national missile defense that could be deployed on a very short timeline. Both the Army and Air Force proposed architectural options to be developed as a plan of that study. The team estimated time scales of approximately four years to deployment and described several opportunities and associated challenges to deploy an interim NMD capability to deal with rudimentary Third World threats to the United States.

Early in 1995, the Department of Defense (DOD) had developed a set of National Missile Defense program options. These included an enhanced baseline development effort, an emergency response system, and an enhanced NMD technology program. The enhanced NMD baseline program became the Department's "3 plus 3" program. The emergency response system included the Air Force and Army Options.

The U.S. Air Force Emergency Response Architecture

The Air Force recommendation consisted of an early deployment option, should a national emergency require fielding a NMD system before the 2003 timeframe. Using the existing Minuteman intercontinental ballistic missile infrastructure, the architecture would deploy twenty Minuteman missiles equipped with kinetic energy kill vehicles in existing silos at Grand Forks AFB, North Dakota.[4] A network of upgraded early warning radars would support the interceptors. The Air Force projected that such an architecture would cost about $2.5 billion and could have been deployed in four years.

The U.S. Army Emergency Response Architecture

The Army responded to the emergency deployment challenge by proposing a booster that combined existing commercial booster stages to launch the kill vehicle. This kill vehicle was already under development. In order to enhance radar coverage, the Army proposed to augment early warning radars and utilize a ground-based radar (GBR) that used technology adapted from the theater missile defense Theater High Altitude Area Defense (THAAD) GBR.[5] The Army estimate included the development, testing, production and fielding of a system that could have been operational in slightly more than four years. The Army proposal basically accelerated an architecture similar to the Defense Department's "3 plus 3" NMD.

The Department of Defense's "3 plus 3" National Missile Defense Program

In response to the evolving ballistic missile threat the NMD program was elevated from a technology development effort to a Deployment Readiness Program. A Joint Program Office was established under BMDO with a charter to develop a NMD system for possible future deployment. The Department of Defense also designated NMD as a Major Defense Acquisition Program (MDAP) to ensure it received an appropriate level of management attention and oversight.[6]

The mission of the NMD system was to defend against an ICBM attack consisting of several missiles from a rogue nation or a very small, accidental launch from more nuclear capable states. The system development was scheduled for completion within three years with an integrated test in 1999, to demonstrate the NMD system's capabilities. The decision to deploy the system was to be deferred until after a successful demonstration and validation of a threat. If a decision to deploy had been taken in 2000, additional funding would have been provided to achieve operational capability in another three years, by 2003.[7]

The Ballistic Missile Defense Organization designed three NMD architectures named Capability 1, 2, and 3. The Capability 1 (C1) architecture was a single site defense from Grand Forks, North Dakota. It would have a deployment of 20 interceptor missiles, with an X-band ground based radar located within 150 kilometers of them.[8] The Capability 2 architecture would build on the existing C1 system, and add a further 80 interceptors to reach a total of 100 interceptors. The space-based infrared satellite, SBIRS-Low cold body tracking satellite would be incorporated into the system which would facilitate earlier launch of the interceptors. The time frame for the upgrade from the C1 to C2 system was around 2010. The Capability 3 architecture would add a further 100 interceptors or more, figures mentioned are in the range of 200 to 250. The interceptors would be spread around two or more sites.

The C1 and C2 interceptor numbers were combined in a statement by the Undersecretary of Defense for Policy with the announcement that any deployment of interceptors would be an initial 100.[9] This alteration was founded on the assumption that the 20 interceptors would provide protection against a threat of around 5 missiles. This would not have provided enough defense against a possible North Korean missile arsenal, so the NMD C1 system would have had to be expanded prior to the 2010 upgrade. Also, in order to meet the requirement of defending the entire United States the Department of Defense began examining a deployment option in Alaska for the interceptors. In the revised plan, 20 interceptors would be deployed in 2005, and the additional 80 would be in place by 2007. The second site would be Grand Forks, North Dakota and would add a further 100 interceptors and be deployed in around 2010-2011.

The National Missile Defense Exoatmospheric Kill Vehicle

The exoatmospheric kill vehicle (EKV) is the interceptor that kills the threat ballistic missile in space. The relevance of National Missile Defense with regard to the militarization and weaponization of space is that the intercept occurs in space. The NMD system incorporates a space weapon, that is to say a weapon with its intended destination confined to the realm of space. The EKV contains sensors to see the target array, a navigation system that plots the course of the vehicle, software to control the flight of the booster, computer processors, and advanced algorithms in order to discriminate the target array and to plot the intercept course and fuel for maneuvering the EKV.[10]

The first generation of EKV has only a passive visual/infrared sensor system which enables it to view the target array in two dimensions. They are equipped with a laser ring gyro inertial measurement system that allows the interceptors to be maintained in a dormant state prior to launch. The intercept control systems can be activated within seconds. The targeting information and a current star map is loaded into the EKV's memory, which guides the interceptors towards their preliminary aim points within a brief time.[11] The EKV does not have an onboard capability to determine the range to the target. The target array appears as points of light of varying intensities and frequencies projected on a horizontal plane. If the intended target is sunlit, the EKV will monitor the target array by means of its visible light sensor system. The two infrared sensors would scan for targets that may be located in dark areas. This would be done by detecting the medium- and long-wave infrared emissions from the target array using medium- and long-range focal plane arrays, each composed of 65,536 pixels.[12] These pixels will be hit by infrared inputs that are filtered to determine the target array objects. The initial EKV will only be able to determine the distance from the target from information provided by the ground-controlled radar tracking.

The closer the EKV comes to the target array, the more the pixels become illuminated. In its final approach the target array will bloom and activate a large amount of pixels and the image of the target will fill the sensor's telescope.[13] The EKV would then proceed to align its path to make an intercept. The ability to divert its path to intercept is the critical determinant in the effectiveness of the kill vehicle. The kill vehicle maneuvers on three axes. The burnout velocity provides most of its motion. To move on the vertical and horizontal axes the onboard thrusters have to be operated; this speed is called the divert velocity. Unlike endoatmospheric interceptors, exoatmospheric kill vehicles are unable to make sharp turning maneuvers. Maneuvering in space permits only minor turning movements. The amount of fuel the kill vehicle carries to operate the thrusters limits its divert capability. The fuel amount is a trade-off, carrying extra fuel adds mass to the payload that could slow divert velocity, but less fuel reduces thrust time, and reduces maneuver capability.[14]

The National Missile Defense Act of 1999

The National Missile Defense Act of 1999 was passed with a veto-proof majority. The 1999 NMD Act contained the following language:

> It is the policy of the United States to deploy as soon as technologically possible an effective National Missile Defense system capable of defending the territory of the United States against limited ballistic missile attack (whether accidental or deliberate) with funding subject to the annual authorization of appropriations and annual appropriation funds for National Missile Defense.[15]

This majority was achievable with the publication in July 1998 of the Rumsfeld Commission to Assess the Ballistic Missile Threat to the Untied States and its finding that the United States could face a ballistic missile threat from a third world country within the next five years with little or no warning, which was sooner than the National Intelligence Estimate had estimated. The report concluded that countries such as Iran and Iraq had been able to obtain technological assistance from other rogue countries and from industrialized countries with relaxed export controls. The CIA reaffirmed its assessment that no hostile country, with the possible exception of North Korea could acquire an ICBM before 2010. However, this was undermined in August 1998 when U.S. analysts were surprised by North Korea's test firing of a two-stage ballistic missile called Taepo Dong 1. The new exoatmospheric missile overflew northern Japan and landed in the Pacific Ocean and traveled approximately 1,500km.[16] Prior to this North Korea had only successfully test fired a single-stage missile. U.S. intelligence had been anticipating the Taepo Dong 1 test, but they had believed that North Korea would have had a harder time in developing a two-stage vehicle.[17] The North Koreans subsequently claimed that they had been launching a satellite into orbit, which was confirmed by the Pentagon and State Department officials who announced that it was an attempt to launch a satellite with a three-stage Taepo Dong 1 booster.[18] The implication of the attempted launch is that if North Korea can orbit a satellite they can build an ICBM.

However, a failed National Missile Defense test in July 2000 led President Clinton to postpone making the deployment decision envisioned in the "3 plus 3" plan. The decision effectively put the deployment decision for a National Missile Defense off until after the presidential elections in September 2000. The emergence of President George W. Bush in the elections meant that the newly elected Republican president would decide the future course of the national missile defense program.

President Clinton's Military Space Policy

The National Space Policy (Presidential Directive 7) was announced on September 19, 1996. One of the most significant aspects of this policy was the convergence of some defense space programs with civil space programs, with the promotion of the Pentagon's utilization of commercial services.[19] The rationale behind this shift to commercial services, especially in the realm of communications is better efficiency and cost savings. Governmental investment in the development of the infrastructure is being transferred to the exploitation of commercial services and products, which are frequently developing at a quicker pace than the public sector.

The National Space Policy acknowledged for the first time the existence of U.S. photoreconnaissance satellites having a near real-time capability for intelligence collection, defense planning, and military operations. The existence of the National Reconnaissance Office and senior officer positions were also declassified. This began a process of eliminating excessive classification of U.S. military space policy and organizations.[20] The space policy directed the defense and intelligence sectors to work closer together. The Secretary of Defense was given authority to propose modifications to intelligence-gathering satellites and develop and operate Defense Department satellites if intelligence space systems could not provide the necessary intelligence support. The National Space Policy adopted many of the lessons learned from Desert Storm, and recognized that space systems had become a critical tool.

The Military Space Organizational Changes under the Clinton Administration

There were a number of significant changes to the organizations responsible for conducting military space operations. The Space Warfare Center (SWC) under the auspices of the Air Force Space Command had its range of tasks increased during the mid-1990s. The SWC's tasks ranged from identifying the merits of futuristic ideas to turning quick reaction space systems into operational units.[21] The SWC now parallels the work of other military warfare centers. It was made responsible for integrating air and space operations. This included developing space tactics, crew manuals, and training courses for Air Force personnel who are involved in utilizing space-derived information to increase mission effectiveness. The SWC will develop and publish a Multi-Command Manual (MCM) for space. This would become a primary information source on space systems" capabilities, including tactics and characteristics that need to be considered when planning a mission. The expanded role of the SWC includes the Space Battlelab, which is the central clearing house for military space ideas. The battlelab based at Schriever Air Force Base focuses on innovative space operations and logistics concepts and tests the concept in operational situations. This evaluation relies on modeling and simulation, along with field-level prototyping and trials. One project considered by the lab was using unmanned aerial vehicles as data relay platforms during satellite launches and early on-orbit operations. Currently, advanced range information aircraft must provide the necessary command and control links where communication or sensor gaps exist.[22]

The SWC also had the 17th Test Squadron added, which has the responsibility for all space forces operational testing and evaluation. The SWC already is home to the 576th Flight Test Squadron which contains the ICBM Minuteman and Peacekeeper flight tests. In addition, modeling and simulation analysis that creates computer models of space systems, both current and future became part of the SWC. They are physics-based models that for example, show where a satellite will be on a certain occasion to ensure a sensor can view a target.[23] Also, the unit produces models for space wargames in multiservice exercises.

One such event was the Pentagon's "Title 10" war game examining military space. "Title 10" was set in the period 2010-2017 and demonstrated the value of advanced systems, such as rapid-response space planes and on-orbit radar constellations. The previous war games conducted by the U.S. Air Force, "Global Engagement V" and the Navy's "Global 2000" war games underscored the importance of a deep-look, rapid strike capability to locate and destroy time-sensitive targets in enemy territory.[24] Prior to this, there had been war games that had included space assets, but Global Engagement V and Global 2000 were more realistic in the sense that the teams had not been given a large number of space-strike capabilities. The insights from these war games were that the United States requires a capability to rapidly reconstitute its national security space platforms. If an adversary knocks out of commission 6, 8, or 10 satellites, these need to be replaced. During Global Engagement V they were replaced by microsatellites which were put into orbit by space orbiting vehicles. The timing signals from the Global Positioning System that are vital to network-centric warfare require multiple platforms and systems in order to synchronize actions. This has led to concerns regarding GPS vulnerability.[25] A further insight was that the United States does not have a current or programmed means to quickly strike important, time-sensitive targets deep inside a country.

The vulnerability of satellites to laser attack has led the U.S. Air Force Research Laboratory's Space Vehicles Directorate to develop sensors that are designed to detect intentional interference with a satellite. The ground crews would then be alerted to such an attack and would be able to take action. The systems would fly onboard military and civilian satellites, detecting, identifying, characterizing, and reporting any radio frequency and laser interference with U.S. and allied spacecraft.[26] This would enable the ground crews to understand the impact any attack might have on the satellites' mission and be able to predict any degradation in performance of the satellite.

The outcomes of these two games were incorporated into "Title 10," the first national-scale war game committed to military space held in January 2001, called the Schriever 2001 war game. The focus of

Schriever 2001 was to explore the requirements for space control, and space force application requirements to support an expeditionary aerospace force, along with ongoing joint-service and government agency needs. In addition to this was to explore possible countermeasures to an adversary's space and intelligence, surveillance, and reconnaissance (ISR) capabilities, and to evaluate an enemy's actions that could deny the U.S and its allies space assets.[27] The participants were high-level current and former military and government officials, along with a "commercial cell" familiar with the space industry.

The war game was set in 2017 and began with a fictional scenario. A large space-capable "near peer" country (Red) massed its forces near the border of a nation (Brown) which called on the United States and its allies (Blue) for protection. The Blue team found it difficult to intervene due to political and economic complications. There were two parallel games played. The Blue team was equipped with a force that the U.S. Air Force could expect to have in 2017 if it continued on its current programd course, buying airplanes and weapons and slowly building its space capabilities. The second Blue team was considered a robust force that would be in place if more funds were committed to developing and fielding a strong air and space presence.[28] These two teams were pitted against the Red teams that had the same assets, enabling a comparison to be made between the effectiveness of the programmed or robust Blue forces. One of the lessons learned was the deterrent capabilities of space assets. Adversaries are less likely to mount a surprise attack if they are aware that their movements are being monitored.[29]

The Clinton Administration's Second Term

In the second term of the Clinton administration, Secretary of Defense William Cohen issued a new Defense Directive to replace the 1987 version, on 9 July 1999. The new directive did not differ significantly from existing policies announced in previous documents. It reaffirmed the importance of military space to U.S. military strategy and the requirement of achieving information superiority.[30] Of particular note, the directive stated that any interference with U.S. space systems would be viewed as an infringement on U.S. sovereign rights, and the U.S. may take appropriate self-defense measures, including the use of force.[31] The directive therefore subtly added to the capability to control space outlined in the 1987 version with the provision "if directed." In addition to this, the directive proposed "the ability to perform space force applications in the future could add a new dimension to U.S. military power."[32]

The directive defines space force applications as: "Combat operations in, through, and from space to influence the course and outcome of conflict. The force application mission area includes: ballistic missile defense and force projection."[33]

Force application from space was included as a long range planning objective and was specifically referred to. The aim was to "explore force application concepts, doctrine, and technologies consistent with Presidential policy as well as U.S. and applicable international law."[34]

The Clinton administration viewed military space with a fairly low priority. Military space budgets were relatively low during the two administration periods. Also, the National Space Council (which had enabled space issues to receive executive-level attention) was terminated. It was replaced by the National Science and Technology Council which has responsibility for space, along with other science and technology policy matters. Within the Pentagon, the deputy undersecretary of defense for space which coordinated space policy and procurement matters was dissolved. This was added to the C^3I office in the Pentagon. The office of national space architect was reformulated into the national security space architect. The Clinton administration had lowered the profile of defense space by merging the individual organizations that had been established into other nonspecific bureaucracies.[35]

President Bush and Missile Defense

President Bush in a speech at the National Defense University announced the administration's policy toward ballistic missile defense for the first time since the presidential elections. The following quote highlights the central thrust and the rationale for national missile defense:

> Today's most urgent threat stems not from thousands of ballistic missiles in the Soviet hands, but from a small number of missiles in the hands of these states, states for whom terror and blackmail are a way of life. I asked the Secretary of Defense Rumsfeld to examine the available technologies and basing modes for effective missile defenses that could protect the United States, our deployed forces, our friends and our allies.[36]

The speech announced a widening of possible national missile defense architectures beyond the former architecture developed under the previous administration. This also includes covering friends and allies. The national missile defense architectural design will focus initially on defending the United States. The "national" has been omitted from National Missile Defense, to Missile Defense to emphasize the international aspect of the new administration's ballistic missile defense policy. The Ballistic Missile Defense Organization was given a new name on 1 January 2002 to Missile Defense Agency.

The United States Withdrawal from the ABM Treaty

On 13 December 2001 the United States provided formal notification that it was withdrawing from the ABM Treaty as set out under Article XV of the Treaty which permits a six month notice of withdrawal. The rationale behind the withdrawal of the ABM Treaty is the altered strategic environment since signing in 1972 especially with regard to the United States' relationship with Russia. Missile defense proponents argue that the present threat environment includes rogue states that are acquiring increasingly longer-range ballistic missiles as an instrument of coercion and blackmail against the United States and its allies.[37] The ABM Treaty prohibits the United States from defending its national territory from ballistic missile attack, and the United States considers defending against these threats to be imperative. The rationale also cites the ABM Treaty's limitations on the United States cooperating with allies with regard to developing missile defenses and its intention to undertake such cooperation in the future.

The United States' withdrawal from the ABM Treaty has important ramifications for the weaponization of space. It removes the international treaty which prohibited the deployment of space-based weapons and the restrictions on nationwide missile defenses. This move paves the way for possible deployment of the space-based laser as discussed in a previous chapter. It also removes any restrictions from the United States' plan to develop a nationwide missile defense system. This missile defense system will incorporate space-based assets and indeed an exoatmospheric interceptor, which will see the further militarization and weaponization of space.

Attempts for Arms Control in Space

There have been recent efforts in the United National Conference on Disarmament to place restrictions on the types of weapons being placed in space. The move was to expand the 1967 Outer Space Treaty to ban all types of weapons. The talks were known as PAROS, the "prevention of an arms race in outer space." The talks have been stalled due mainly to the objections of the United States.[38] In November 2000 the United States, Israel, and Micronesia refused to vote for a UN resolution citing the need for steps to prevent the arming of space. It is not surprising that the United States does not want to place limits on its plans to use space for missile defense and space control purposes. Russia and China have been proponents pushing for such a treaty. Indeed, Russia has been advancing the establishment of an ad hoc negotiating structure as a preliminary step.[39]

The Commission to Assess U.S. National Security Space Management and Organization

The Space Commission report called for a top-down realignment of the Department of Defense's space infrastructure. The panel recommended that the Air Force be given "Title 10" responsibility for organizing, training, and equipping space forces, and making the U.S. Air Force the nation's executive agent for space.[40] However, there were provisions for initiating a transfer to a separate space corps "as soon as practicable" which is being widely interpreted as 5-10 years from now. The Commission cited the Navy's nuclear branch as a model the Air Force might want to consider for its space forces. This would mean that space would have an independent promotion system and funding. Admiral David E. Jeremiah, the panel's new chairman after Rumsfeld stepped down, indicated that the commissioners preferred an initial arrangement to keep the space experts within the basic Air Force. The commission did not advocate the creation of a near-independent service, such as the Marine Corps, but preferred the former Army Air Corps structure.[41]

The Space Commission and the implementation of some of its recommendations have resulted in some significant organizational changes in the realm of military space policy. The importance of the Space Commission report that was released in January 2001 was signified with the chairman Donald Rumsfeld's elevation to the position of Secretary of Defense. A National Security Space Architecture (NSSA) office has been established to examine near-term transformation issues. Prior to this the Pentagon's space architecture was focused on defining communications satellites and other architectures with a time frame for implementation in around 15-25 years time. Army Brigadier Stephen Ferrell is the NSSA chief, with responsibility to integrate the needs of the Army, Air Force, Navy, Marine Corps, and other agencies. Associated with this a Defense Space Acquisition Board was established to reduce a program's milestone approval process from a year to a matter of weeks. In addition to this a Program Executive Officer for U.S. Air Force space programs was designated, who would report directly to the Defense Department Executive Agent for Space on space system acquisition issues.

A Directorate of National Security Space Integration was created with the responsibility of combining the best practices of the black (secret) and white (open) space worlds to improve the integration processes.[42] This is run by Brigadier General Michael Hamel, and contains around 25-30 personnel. This office is intended to support Peter Teets, who is the Director of the NRO and Air Force undersecretary and is the executive agent for space. It is also tasked with coordinating initiatives ranging from communication between the military and intelligence agencies, to planning, policy development, and budget synchronization.[43] In particular on policy, the organization will be involved on export controls and international cooperation on space matters.

A significant boost for military space was the appointment of a four-star commander for Air Force Space Command. Previously, one general simultaneously headed U.S. Space Command, Air Force Space Command and the North American Aerospace Defense Command. The U.S. Navy is increasing the importance of Naval Space Command. The Navy is forming a Network Warfare Command (NetWarCom) under a three-star admiral. As part of the changes, Naval Space Command will be called Naval Network and Space Operations Command under the NetWarCom. Also it will move offices to Norfolk, Virginia. These changes should make the personnel and control center more integral in real time operations and a stronger advocate for naval space research and technology development.[44]

Indeed the Space Commission chaired by Rumsfeld added impetus to the case for developing an antisatellite capability. The prestige of the Space Commission was further enhanced when its chair was subsequently given the post of Secretary of Defense. The Commission highlighted the enormous vulnerability of the United States' space assets to possible attack and strongly emphasized the United States' dependency on these assets.[45] The strong emphasis and the political clout of the Space Commission highlighted one of the major rationales for developing an ASAT capability. This rationale was the requirement to be able to disrupt an adversary's space capabilities and perhaps to deter an adversary from undertaking an attack on U.S. space capabilities, but at least to be able to conduct some form of retaliatory attack on an adversary's space assets. To ensure that the adversary is unable to glean advantage with its space assets during a conflict, while the United States had lost its military space capabilities, the Space Commission states that "the U.S. must have the capabilities to defend its space assets against hostile acts and to negate the hostile use of space against U.S interests."⁴⁶ It goes on to advocate a U.S. ASAT requirement by declaring that "the U.S. will require means of negating satellite threats, whether temporary and reversible or physically destructive."⁴⁷

The Space Commission provided strong political pressure for the United States to develop an antisatellite capability. The report provided a timely reminder of the United States' vulnerability to space threats and its current lack of ability to deal with such threats. The advocacy of an antisatellite capability by the panel of experts contributed greatly to the antisatellite debate, and provided an impetus to the politics of acquiring such a capability.

Space-based Missile Defenses

The restructuring of the Space-based Laser (SBL) program led to the cancellation of the Integrated Flight Experiment that was to develop a demonstration experimental laser in space by 2012.⁴⁸ The restructuring of the program is part of a wider Missile Defense Agency approach that is following a capability-based approach. This approach is to identify affordable operational concepts and to focus on early contributions to midcourse and terminal systems. The focus of the SBL program is on key technologies supporting a future system, with no specific focus date. The restructuring of the program could lead to a redefined experimental SBL ahead of the original 2012 schedule.⁴⁹

The space-based hit-to-kill program that was cancelled in 1993 was resurrected in the Fiscal 2002 budget.⁵⁰ The initial funds are to assess the technology available for a system with a midterm goal of an in-orbit experimental system by 2005-2006. The space-based experiment will differ from the original Brilliant Pebbles concept planned in the early 1990s, in several ways. The current interceptor would be launched in close succession with the target and would have no station-keeping capability.⁵¹ It would be essentially a one-time event and subsequently any follow-on tests would require new hardware. The Brilliant Pebbles concept envisioned placing space-based interceptors (capable of performing multiple engagements), in orbit for around eighteen months.

Although research in the area has been absent for a decade, through developments in the commercial sector and in other defense programs it is believed that the Brilliant Pebbles concept could be updated for boost-phase intercept from space.⁵² The efficiency of small pump-fed engines and the impulse of solid axial engines have improved. These were a major limitation in the acceleration and maximum velocity of Brilliant Pebbles. An interceptor with a 10g acceleration and a maximum speed of 10kms^{-1} would have a range of around 400km to intercept a target that requires 90 seconds to accelerate.[53] An interceptor with 20g acceleration could cover a range of 800km.[54]

Space-based Weapons against Terrestrial Targets

The notion of space force applications mentioned in the Space Policy directive in 1999 included force projection from space. Prior to this the U.S Space Command's Long Range Plan (LRP) explored the concept of force application from space, but placed this caveat that "at present, the notion of space weapons in space is not consistent with U.S. National Policy. Planning for this possibility is the purpose of this plan should our civilian leadership later decide that the application of force from space is in our national interest."[55] The LRP outlines an aspiration that by 2020 concepts such as Conventional Ballistic Missiles Common Aero Vehicle, Space-Based Platform, and Space Operations Vehicle will provide on-demand precision engagement from space.[56] The Common Aero Vehicle (CAV) would be a satellite deployed in low-earth orbit and would slow from orbital speeds to dispense conventional munitions.[57] Another option for the CAV is to use air-launched suborbital missiles or ICBMs to deliver it. The slowing down of the CAV relinquishes the ability to hit targets from orbital speeds.[58] These concepts will provide the ability to attack a number of targets, be they fixed, relocatable, or moving high-value targets, nearly instantaneously.

The Defense Science Board, a senior advisory panel to the U.S. Department of Defense recommended that the Pentagon consider researching, developing, and fielding a series of new capabilities. Of these new capabilities those pertaining to space weapons are a two-stage intercontinental ballistic missile-launched precision weapon using GPS for guidance using kinetic energy or conventional projectiles.[59] A further recommendation includes a constellation of orbiting vehicles containing rods of heavy material in highly elliptical orbits to reenter the atmosphere, and hit targets at speeds of 10,000fts^{-1}.[60] The concept of kinetic energy striking from space to terrestrial targets uses the kinetic energy obtained from the weapon's high velocity (around 5 to 11kms^{-1}). However, to reach velocities in the region of 11kms^{-1} the weapons would have to be in orbits at an altitude of more than 40,000km and would require around five hours to hit the earth's surface.[61] Therefore a sacrifice in the velocity of the projectile would have to be made in order to make the weapon more time responsive. Lowering the orbit of the weapon to an orbit of 500 miles (926 km) would reduce the velocity to 5kms^{-1}, and it could strike in less than 12 minutes.[62]

These weapons would be very difficult to defend against due to the high velocity achieved from their operation from space and may be useful against heavily defended targets. The United States has other methods of power projection so these weapons may be of only limited interest, but another country that seeks global power projection and does not wish to emulate the U.S. defense investment could have an interest in them.[63] One design suggested for use as a kinetic energy project is a thin, heavy metal rod one or two meters long.[64] The rods would have to remain symmetrical and be delivered with a zero angle of attack to avoid any tendency to fly. The range of targets that could be hit using metal rods would include tall buildings, missile silos, and hardened aircraft shelters, but not runways, deeply buried bunkers, bridges, and long low buildings.[65]

Other targets that have been mentioned as being susceptible are surface ships including aircraft carriers,[66] although this has been caveated with the proviso as long as the ship is unable to move too far unpredictably in the few seconds it takes the weapon to reenter the atmosphere.[67] This arises since targeting adjustments are possible during the weapon's flight outside the atmosphere, along with small changes up to a few tens of seconds before impact.[68] Other designs are to fit the rod with an ultrahard penetrator with an explosive warhead, or with a warhead that fragments upon impact. The penetrator approach enables the weapon to penetrate deeper than eroding rods by detonating at a preset depth using the time from initial impact, or when it reaches an area of low resistance such as a room. However, the materials that are hard enough to remain intact during the penetration phase are still under investigation.[69]

A concept that although not space-based would indeed traverse through space is transatmospheric vehicles (TAV). These weapons would be launched on demand and would be designed as reusable vehicles to put payloads in orbit and deliver them anywhere in the world within hours.[70] The platform that holds most promise to launch the TAVs is an aircraft. The aircraft effectively acts as the first stage of the TAV.[71] A similar concept was used with the miniature homing vehicle in an antisatellite role aboard a modified F-15. A limitation associated with using an aircraft as a basing mode for the TAV is that the weight is restricted to the capacity of the aircraft. However a Phillips Laboratory Military TAV Technical Requirements Document specifies a desired payload of 1,000 pounds[72] that is within the capability of aircraft platforms.

The advantage of TAVs over space-based weapons is the ambiguity of the basing of the weapon. The ambiguity lies in the fact that it is not based in space. The weapon is launched when required and is not permanently based in space. This political nuance gives TAVs an advantage over space-based platforms despite the fact that it does strike from space with the same power as space-based weapons since it travels at orbital speeds.[73] In addition to this politically beneficial factor, TAVs have an advantage since they could be launched from aircraft and could operate at a lower cost than space-based equivalents. The launch on demand system TAV would face the same degree of difficulty to intercept as space-based weapons in orbit and could possibly be more difficult than those on an orbital track in space.[74]

The Role of Military Space in Recent Conflicts

Since the Persian Gulf War there has been a concentration on getting space-derived information to troops in the field. The U.S. Space Command commanders in chief have focused on operationalizing space, that is transferring information from satellites to frontline commanders to enable them to leverage space assets more efficiently.[75]

The Yugoslavia Campaign

Intensive space reconnaissance missions were undertaken to lead NATO targeting operations and bomb damage assessments of the strikes against Yugoslavia in 1999. Imagery analysts were under pressure to provide imagery to strike planners and the air crews undertaking the missions. There had been concern prior to the air strikes whether a late 1996 reorganization of imaging operations had left the intelligence community with sufficient analysts to sustain such a large operation.[76] Another area of concern was whether the imagery intelligence community would be able to maintain critical watch over priority areas such as North Korea, Iraq, China, India, and Pakistan whilst attention was being focused on Yugoslavia.

The National Reconnaissance Office's three advanced KH-11 visible/infrared electro-optical satellites were used to image refugee lines of individuals from orbits of 170 x 629 miles altitude. Each of them flew over the conflict area twice daily and provided some slant-range imagery on passes to the east and west of the region. The infrared sensors were also capable of detecting hundreds of tiny camp fires from displaced persons in southern Yugoslavia. It was speculated that three other highly secret, smaller NRO imaging satellites may have been involved in the operation.[77] There was an unprecedented level of space systems involved in multiple areas of the NATO action, which indicates the level of air and space integration. The GPS constellation provided critical navigation data to air crews and specific precision-guided weapons to their target. The standard GPS system is comprised of twenty-four satellites, but the system was enhanced to twenty-seven satellites.[78]

During the operations NATO strike planners and space officers coordinated their planning to ensure that precision-guided weapons were dropped when GPS satellites were in the best position to direct the weapons onto the specified targets. The relative positions of specific GPS satellites and strike coordinates and timing enabled a calculation to ascertain the best navigation data to the aircraft or weapon. The Kosovo campaign was the first time the United States used the Joint Direct Attack Munitions, an all-weather precision-guided bomb, that costs around $21,000 as opposed to the $1 million Tomahawk cruise missiles.[79] The GPS receiver, depending upon the type used in the aircraft or weapon, could be receiving data from as many as eleven GPS satellites. The 2nd Space Operations Squadron under the 50th Operations Group at Schriever, Air Force Base, Colorado manages the GPS navigation payloads for strike support.

Over fifty U.S. and European satellites were involved in NATO coordination, intelligence and strike operations. In planning and executing the attacks there were more than 15-20 different U.S. and European types of space systems. The U.S. Air Force Space Command classified the previously public orbital data on U.S. military satellites. This denied the open flow of data to trained analysts in Yugoslavia which could have used the information to determine the overflight times of the satellites involved in strike and intelligence support, which could have provided clues to the timing of tactical operations.[80] The Lacrosse imaging radar system provided pre-strike intelligence and post-strike damage assessment of targets in forest terrain, which is sometimes obscured from visible or infrared systems by inclement weather. Two of the Lacrosse imaging radar satellites flew over the Yugoslav/Kosovo area twice daily.[81] These satellites provided 1-3 feet resolutions that enabled damage assessments during the night and in all weather conditions. It also enabled Serbian armour or mobile SA-6 surface-to-air missile systems to be targeted, even if they were concealed in forested areas.

The weather conditions in Eastern Europe were such that local conditions were frequently changing. This made weather satellite operations vital to strike planning and the timing of reconnaissance imaging operations. There were ten U.S. and European weather satellites that provided imagery which aided these operations. These included four U.S. Air

Force Defense Meteorological Satellite Program satellites in 500-mile polar orbits. These could provide weather image resolutions as small as around 1,000 feet. Four National Oceanic and Atmospheric Administration polar-orbit weather satellites, NOAA-10, 12, 14, and 15 also assisted in this capacity. NOAA-12 and 15 were positioned on opposite sides of the same ground tracks, which enabled them to overfly the same areas an hour apart, which provided information on short-term changes.[82]

The Events in Afghanistan

Space systems provided the collection, processing, and dissemination of time-critical information to help forces locate enemy forces. It enabled the precise targeting of an ethereal enemy, rapid and effective air strikes, and minimized allied casualties. In particular, near-real time video from Predator UAVs was relayed by communications satellites to enable targets to be identified and attacked on the ground.[83] The Army coordinated with national agencies to produce 3-D "fly throughs" of terrains before troops and aviators approach that area. Space control assets have been in operation to deny Al Qaeda and Taliban forces satellite communications.

The U.S. Naval Space Command provided communications support which has been vital to the Marine Corps, but the Remote-sensing Information Center has provided the Marines with multispectral satellite (MSS) imagery of shorelines along the Arabian Sea and the Straits of Hormuz.[84] This assists the Marine Corps with information in which intelligence analysts can determine landing zones or Al Qaeda fighters in concealed locations. The command has also contributed to the United States homeland defense, with the MSS imagery of around twenty-eight ports to assist with port security.

During the campaign in Afghanistan, the U.S. launched a fourth Milstar defense communications satellite to make the network for rapidly moving communications to military forces more robust. The three operational Milstars had seen a 10-15 percent increase in theater operations traffic since the September 11 attack. The fourth satellite is expected to improve flexibility for Army, Special Ops Forces, and other users by permitting relays of high-priority information, including imagery, around the Earth via satellite crosslinks. Space to Earth bounces through ground stations are no longer necessary.[85] Milstar has been used during the campaign to route updated Tomahawk targeting data to U.S. Naval carrier battle groups in the Central Asian theater.

U.S. Space Command and its Air Force, Army, and Navy components have refined military satellite communications procedures and augmented capacity with commercial resources as required.[86] The controllers at Schriever Air Force Base Colorado, finetuned the GPS satellites to ensure the premium navigation information was available for near-precision GPS-aided weapon deliveries. Increasingly efficient use of bandwidth when downloading weather satellite data has provided more frequent updates to terrestrial and space weather forecasts. These are critical to the success of air strikes and special forces missions.

Conclusion

The issue of military space has significantly come to the fore in the period following the end of the Cold War. The issue that has made the most impact with regard to the increasing militarizing and weaponizing of space is missile defense. This issue in particular was prevalent in both terms of the Clinton administration. There were a series of National Missile Defense Acts initiatives from Congress in an attempt to develop a system to counter a limited ballistic missile attack on the United States. There was stringent opposition to these proposals from President Clinton until out of political expediency prior to the 1996 presidential elections the requirement for a missile defense system was accepted, in the form of the "3 plus 3" program. This missile defense architecture envisioned the use of an exoatmospheric interceptor, that impacts on the threat ballistic missile in space. The lack of success during the initial testing period led the decision of deployment during the Clinton presidency to be deferred to President Bush. The Bush administration announced a fundamental review of national missile defense, and removed the "national" from its terminology to announce a widening of missile defense, to include allies. The withdrawal from the ABM Treaty signaled the serious intent of the new administration to develop a nationwide missile defense system some time in the immediate future.

Military space policy during the Clinton administration received fairly low prioritization. The 1996 National Space Policy did not differ significantly from the previous space policy save for the declassifying of the National Reconnaissance Office. The new directive in 1999 subtly added the proviso "if directed" to the capability to control space, along with the exploration of the concept of space force application. The directive did not specifically announce the development of space control and space force application, merely the research and development of these concepts. The termination of the National Space Council and the dissolving of the deputy undersecretary of defense for space lowered the profile of defense space by merging the previous organizations into nonspecific bureaucracies.

The profile of military space policy under the Bush Administration was raised considerably with the announcement of Donald Rumsfeld as Secretary of Defense, who was previously chairman of the Space Commission. Many of the organizational changes that were recommended by the Space Commission were implemented to increase the influence of military space within the policy-making apparatus. These included the appointment of a four-star commander for Air Force Space Command, the creation of a directorate of national security space integration and a national security space architecture office. It is too early in the Bush administration to determine the impact these offices will have in advancing military space issues.

The issue of missile defense has considerable implications for the weaponization of space. The ground-based system for protection of the United States uses an exoatmospheric hit-to-kill interceptor, which is essentially a space weapon. The development of the space-based laser for boost phase missile interception is a space weapon. The space-based kinetic kill weapon system, the follow-on to Brilliant Pebbles, is a space weapon that intends to intercept ballistic missiles in their boost phase. These three weapon systems, all of them space weapons are being developed for the ballistic missile defense role. This is the primary rationale for the weaponization of space.

Notes

1. Thomas Moore, "The Missile Defense Act of 1995: The Senate's Historic Opportunity," *The Heritage Foundation*, August 1, 1995.
2. *The Defend America Act of 1996* (Washington, D.C.: U.S. Government Printing Office, March 1996).
3. Frances Fitzgerald, *Way Out There in the Blue: Reagan, Star Wars and the End of the Cold War* (New York: Simon and Schuster, 2000), 493.
4. Richard D. West, *Near-Term National Missile Defense Options*, House National Security Committee, Subcommittee on Military Research and Military Procurement, June 18, 1996.
5. West.
6. West.
7. West.
8. David R. Tanks, *National Missile Defense: Policy Issues and Technological Capabilities* (Washington, D.C.: Institute for Foreign Policy Analysis, 2000), 4.3.
9. Walter B. Slocombe, quoted in Tanks., 4.7.
10. David. R. Tanks, 3.13-3.14.
11. Tanks, 4.10.

12. Tanks, 3.14.
13. Tanks, 3.15.
14. Tanks, 3.16.
15. *National Missile Defense Act of 1999*, 106th Congress, 1st session, House Resolution (Washington, D.C.: U.S. GPO, 1999), 4.
16. Joseph C. Anselmo, "Missile Test Extends North Korea's Reach," *Aviation Week and Space Technology*, September 7, 1998, 56.
17. Anselmo, 56.
18. David A. Fulghum, "North Korea Space Attempt Verified," *Aviation Week and Space Technology*, September 21, 1998, 30.
19. Steven Lambakis, *On the Edge of Earth: The Future of American Space Power* (Lexington: University Press of Kentucky, 2001), 232.
20. Lambakis, 233.
21. William B. Scott, "Space Warfare Center Aims to Be 'Nellis of Space,'" *Aviation Week and Space Technology*, September 1, 1997, 49.
22. William B. Scott, "USAF Space Battlelab Assessing New Concepts," *Aviation Week and Space Technology*, September 1, 1997, 52.
23. Scott, 50.
24. William B. Scott, "Wargames Zero in on Knotty Milspace Issues," *Aviation Week and Space Technology*, January 29, 2001, 53.
25. Scott, 55.
26. William B. Scott, "New Satellites Sensors Will Detect RF, Laser Attacks," *Aviation Week and Space Technology*, August 2, 1999, 57.
27. William B. Scott, "Wargames Zero in on Knotty Milspace Issues," *Aviation Week and Space Technology*, January 29, 2001, p53-54.
28. William B. Scott, "Wargame: 'Space' Can Deter, Defuse Crisis," *Aviation Week and Space Technology*, February 5, 2001, 40.
29. Scott, 40. and James Kitfield, "The Permanent Frontier," *The National Journal*, March 17, 2001, *http://www.globalsecurity.org/org/news/2001/010317-nj.htm*.
30. Department of Defense Directive, *Space Policy*, Number 3100.10, July 9, 1999.
31. *Space Policy*, 6.
32. *Space Policy*, 3.
33. *Space Policy*, 23.
34. *Space Policy*, 9.
35. Steven Lambakis, 234.
36. President Bush speech at the National Defense University, Washington, D.C., May 1, 2001 quoted in Bhupendra Jasani, "U.S. National Missile Defense and International Security: Blessing or Blight?" *Space Policy*, November 2001, 243.
37. *ABM Treaty Factsheet*, Office of the Press Secretary to the White House, December 13, 2001.
38. Theresa Hitchens, "Rushing to Weaponise the Final Frontier," *Arms Control Today*, September 2001, http://www.armscontrol.org/act/2001_09/hitchenssept01.asp.
39. Hitchens.
40. William B. Scott, "USAF Warned to Bolster or Lose 'Space Force' Franchise," *Aviation Week and Space Technology*, January 29, 2001, 55.
41. Scott, 55.
42. William B. Scott, "Milspace Comes of Age in Fighting Terror," *Aviation Week and Space Technology*, April 8, 2002, 78.
43. Robert Wall, "Space Reformers Juggle War, Acquisition Demands," *Aviation Week and Space Technology*, April 8, 2002, 81.

44. "Navy Shift Elevates Space," *Aviation Week and Space Technology*, April 8, 2002, 88.

45. *The Commission to Assess U.S. National Security Space Management and Organization* (Washington, D.C.: U.S. Government Printing Office), January 11, 2001.

46. *The Commission to Assess U.S. National Security Space Management and Organization*, 13.

47. *The Commission to Assess U.S. National Security Space Management and Organization*, 29.

48. Colonel Ivette Falto-Heck, Sytem Program Director, SBL Project Office, Space and Missiles Systems Center Air Force Space Command, "Space-based Laser (SBL) Requirements for Expermental Missile Defense System," Presentation at *SMi Military Battle Space Conference*, May 29, 2002, 17.

49. Interview with Colonel Ivette Falto-Heck, Sytem Program Director, SBL Project Office, Space and Missiles Systems Center Air Force Space Command, May 29, 2002.

50. Robert Wall, "Space-Based Interceptor Gets New Lease of Life," *Aviation Week and Space Technology*, August 13, 2001 http://www.awstonline.com.

51. Wall.

52. Gregory Canavan, a Los Alamos Laboratory engineer, quoted in Robert Wall, "Space-based Interceptor Gets New Lease of Life," *Aviation Week and Space Technology*, August 13, 2001.

53. Canavan.

54. Canvan.

55. Howell M. Estes, *U.S. Space Command Long Range Plan* (Colorado Springs: Headquarters, U.S. Command, 1998), 65.

56. Estes, 67.

57. William L. Spacy II, Major USAF, "Does the United States Need Space-based Weapons?" *College of Aerospace Doctrine, Research and Education* (Maxwell Air Force Base: Air University Press, 1999), 29.

58. Spacy, 29.

59. Bryan Bender, "U.S. Blueprint for Future Weapons Systems Is Outlined," *Jane's Defense Weekly*, May 26, 1999, 11.

60. Bender, 11.

61. William L. Spacy II, Major USAF, "Does the United States Need Space-based Weapons?" *College of Aerospace Doctrine, Research and Education* (Maxwell Air Force Base: Air University Press, 1999), 26.

62. Spacy, 27.

63. Bob Preston, Dana J. Johnson, Sean J. A. Edwards, Michael Miller and Calvin Shipbaugh, *Space Weapons Earth Wars* (Arlington, VA: RAND, 2002), 40.

64. William L. Spacy, 27.

65. Bob Preston, et al, 40-41.

66. Kenneth Roy "Ship Killers from Low Earth Orbit," *U.S. Naval Institute Proceedings*, October 1997, 40-43.

67. Bob Preston, et al, 40.

68. Bob Preston, et al, 43.

69. William L. Spacy, 27.

70. William L. Spacy, 78.

71. David Gonzales, Mel Eisman, Calvin Shipbaugh, Timothy Bonds, and Anh Tuan Le, *Proceedings of the RAND Project AIR FORCE Workshop on Transatmospheric Vehicles* Rand Report MR-890-AF (Santa Monica, CA: RAND, 1997), 32

72. Gonzales, 14.
73. William L. Spacy, 79.
74. Spacy, 81.
75. William B. Scott, "Cincspace: Focus More on Space Control," *Aviation Week and Space Technology*, November 13, 2000, 80.
76. Craig Covault, "Recon, GPS Operations Critical to NATO Strikes," *Aviation Week and Space Technology*, April 26, 1999, 35.
77. Craig Covault, "Military Space Dominates Air Strikes," *Aviation Week and Space Technology*, March 29, 1999, 32.
78. Craig Covault, "Recon, GPS Operations Critical to NATO Strikes," *Aviation Week and Space Technology*, April 26, 1999, 36.
79. James Kitfield, "The Permanent Frontier," *The National Journal*, March 17, 2001, http://www.globalsecurity.org/org/news/2001/010317-nj.htm.
80. Craig Covault, "Military Space Dominates Air Strikes," *Aviation Week and Space Technology*, March 29, 1999, 31.
81. Covault, 32.
82. Covault, 33.
83. William B. Scott, "Milspace Comes of Age In Fighting Terror," *Aviation Week and Space Technology*, April 8, 2002, 77.
84. Craig Covault, "Naval Space Ops Crucial to Afghan War," *Aviation Week and Space Technology*, April 8, 2002, 86.
85. William B. Scott, "Milstar Ring to Speed Data toward Combat Zones," *Aviation Week and Space Technology*, January 21, 2002, 28.
86. William B. Scott, "Space Enhances War on Terrorists," *Aviation Week and Space Technology*, January 21, 2002, 31.

Conclusion

The development of the United States policy during the Eisenhower administration followed the sanctuary view of space. The focus was on using satellites for reconnaissance purposes combined with a reluctance to countenance the protection of these satellites by means other than international law. Also, a treaty based approach to establish legal overflight of national territories of these reconnaissance satellites followed the sanctuary view of space power. The Air Force aerospace doctrine advanced by General White differed significantly from the sanctuary view but this approach met strong resistance from the Eisenhower administration. During the Kennedy and the subsequent Johnson administration the sanctuary view of space power was maintained with the signature of the Outer Space Treaty which prohibits weapons of mass destruction being placed in orbit. This significantly curtailed the high ground view of space power that sees space as a place from which Earth could be dominated, presumably with the placing of nuclear weapons in space. The policy of using satellites for reconnaissance continued to follow the sanctuary view of space.

The Nixon administration's most significant space policy act was the signing of the ABM treaty in 1972. This placed limits on ballistic missile defenses and hence had enormous implications for the high ground of military space power theory which sees ballistic missile defense in space as an integral aspect of the military utility of the "high ground." Also, the SALT I Treaty for the first time advocated national technical means (reconnaissance satellites) as a means of monitoring arms control agreements. This action followed the sanctuary school of space power, as a means of using space for enhancing arms control, as a way of strengthening strategic stability since it was believed that that no country would cheat if there was a reasonable chance that the other side had a means of verifying whether they were adhering strictly to the terms of the treaty. The Carter administration pursued a similar policy on satellites for national technical means of verification that followed the sanctuary view of space power theory. President Carter faced with potential satellite vulnerability embarked upon a policy of research and development into a possible antisatellite capability. The research and development of such a capability would tend towards a space control view of space power.

However, it can be assumed that the Carter administration was developing an antisatellite capability as a prelude to an antisatellite ban, using an antisatellite capability as a negotiating tool with which to bargain away. However, such arms control measures proved difficult to negotiate and were never realized.

The Reagan administration's space policy was a dramatic departure from the sanctuary school of space power. The development of an antisatellite weapon while maintaining that an antisatellite treaty was undesirable gravitated towards the space control view of space power. Combined with the announcement of the Strategic Defense Initiative, this ASAT policy leaned Reagan's space policy towards the high ground. Indeed, the development of a space-based ballistic missile defense was one of the fundamental tenets of the high ground view of space power. The announcement of a military space doctrine which valued space support, force enhancement, space control, and force application, meant that space policy leaned heavily toward the space control view of space power. However, to summarize, the Reagan administration's space policy can be classified as following the space control view of space power, but with a view toward the future of a high ground view of space, with the research and development of space-based ballistic missile defenses. The Bush administration facing an altered geostrategic environment, especially vis-a-vis the United States and Soviet relationship redirected the SDI mission to the Global Protection Against Limited Strikes mission. This redirection of the SDI mission clearly weakened the notion of space being free from weaponization, as the space architecture envisioned space components for the interception of ballistic missiles. The rationale for using space components for ballistic missile defense was that they could provide a layered approach which would allow multiple early engagements away from defended areas. However, not unlike the original SDI programs the use of space for ballistic missile interception was left primarily a research and development program.

The United States' use of space during the Cold War period could be characterized as following the broad theory of the sanctuary view of space power. The successive administrations from Eisenhower up to Carter envisioned military space as best serving national security interests through the adoption of a sanctuary view of space power. The Reagan administration saw the prevailing view of space power to be questionable. The announcement of the Strategic Defense Initiative ushered in a more robust view of space-borne platforms for ballistic missile interception. This had implications for the sanctuary view of space in that it was effectively replaced by a space control view of space power which sought to research and develop space weapons. It did not see space as a sanctuary from military operations and viewed space as an arena not too dissimilar to land, sea, and air power.

The United States and the Soviet Union's approaches toward ballistic missile defense differed considerably during the period before the signing of the ABM Treaty. The United States was initially enthusiastic towards the development of a ballistic missile defense system, however this enthusiasm disappeared shortly before the signing of the ABM Treaty. The proposed Safeguard system which contained both exoatmospheric and endoatmospheric interceptors, became a pawn in the arms control negotiating process, and with the signing of the ABM Treaty the United States dismantled the Safeguard site. Indeed, the ABM Treaty marked the death knell for ballistic missile defense in the United States, until the Strategic Defense Initiative in 1983. The Soviet Union on the other hand showed continued interest in both ballistic missile defenses before the ABM Treaty and after its signing. Indeed, considerable work was done on the Moscow antiballistic missile site in the period after the signing of the ABM Treaty and their interceptors were enhanced along with their associated radar and tracking facilities. The Soviet Union's approach differs from the United States in that it did not require ballistic missile defense technologies to be proven before the system was deployed. The Soviet ballistic missile defense system was focused around Moscow.

The architectural designs of the United States and the Soviet Union's ballistic missile defense systems were essentially very similar. The two systems incorporated a layered defense with endoatmospheric and exoatmospheric interceptors. The Soviet Union initially focused its program principally on exoatmospheric interception for the Moscow ballistic missile defense system. It is not the case that the two sites" similarities were due to mirror imaging, but that they recognized that a layered missile defense system offered the best means of protection from ballistic missiles. The use of the atmosphere in distinguishing between warheads and decoys provided a rationale for the inclusion of an endoatmospheric interception. The dismantling of the Safeguard site marked a nadir in the United States ballistic missile defense program. The dismantling of the Safeguard site saw the issue more or less disappear from the political scene until the early 1980s. This was in marked contrast with the Soviet Union which continually upgraded and maintained its operational ballistic missile defense system. Indeed, the Moscow site was continually upgraded up to the demise of the Soviet Union in the early 1990s.

The Soviet Union's approach to military space can be categorized as following the broad principle of the sanctuary school of space. The deployment of photoreconnaissance satellites along with ocean surveillance satellites follows the space sanctuary philosophy in that space should be weapon free and that reconnaissance satellites for arms control purposes strengthen the agreements. The Soviet Union developed the Fractional Orbital Bombardment System and its antisatellite capability. These two space weapons demonstrated that within Soviet thinking there were views which followed what could be interpreted as the high ground of space power, which views space as the ultimate arena in which to deploy weapons. The FOBS system would be well suited to the high ground of space view of space with its ability to strike with weapons of mass destruction with an extremely short flight time. The strong emphasis on an ASAT capability follows the high ground view, but combined with the Soviet military strategy in relation to space control, it would tend to demonstrate the space control school of space power, which sees space as another geographical arena from which military operations can be conducted.

Following the SALT II negotiations, the Soviet Union refrained from testing the FOBS. This action indicated that while the Soviet Union flirted with the high ground and space control views of space power they moved away from embracing these philosophies fully. The subsequent actions in relation to military space demonstrated that the sanctuary view of space power was again in the ascendant.

In the early 1990s during the period of transformation of the Soviet Union to the creation of Russia there was a period when it appeared that there might be some form of cooperation on missile defenses with the Untied States. The GPALS missile defense system was initially embraced by President Yeltsin and some progress was made toward creating working groups exploring areas for technological cooperation. A number of high-level group meetings took place and it appeared that thinking toward missile defense in Russia had become favorable. The Clinton administration took office in 1993 and did not share the same attitude with the previous administration regarding missile defenses and it appeared that the offer of cooperation with Russia on missile defenses disappeared. The GPALS system was no longer under consideration and any potential cooperation with Russia had disappeared.

The Russian military recognizing the importance of military space but faced with an austere budget environment was able to embark on a cooperative program with the United States. This project saw the conversion of ICBMs into space launch vehicles. This venture was developed by the Kompleks scientific and technical center and gained financial backing from the company IVK, along with cooperation with Lockheed Martin. The project produced a major source of hard currency. NASA has also purchased goods and services from Russia, much of which is done in relation to the space station. This support and marketing has enabled the Russian aerospace industry to keep afloat in a period which saw an austere budgetary environment.

The People's Republic of China has developed a considerable military space capability. In particular the development of its communications, photoreconnaissance—in different spectral forms such as synthetic aperture radar, electronic intelligence, and navigation satellites—proffer some military space capabilities. It is also actively considering the development of space weapons, in the form of antisatellite weapons. The ASAT capabilities under examination are direct ascent missiles, ground-based lasers and a parasitic satellite with an explosive charge. The reconnaissance capabilities would enable China to monitor Taiwan's defenses to be used for a possible attack if it does not adhere to the principles China has laid out for Taiwan in relation to its status. In addition, the navigation satellites could be enhanced for use in increasing the accuracy of China's ballistic missile capabilities. The PLA has outlined two missions with respect to military space. The first mission is information support and the second is battlefield combating. The first priority is information support which incorporates intelligence, navigation/positioning, and communications.

China has forged a number of international agreements in the realm of space. These are often termed as civilian space ventures, however many of them have dual-use capabilities. To reiterate, the miniaturization satellite technology developed from a British company could be used as part of China's parasitic antisatellite capability. Similarly, the development with Brazil of its CBERS series of satellites assists China in developing photoreconnaissance capabilities. This cooperation has allowed China to develop its military space capabilities considerably quicker than it would have been otherwise able to do so. These international collaborations have enabled China to develop an array of satellites dedicated to military space purposes.

The Soviets rigorously developed a co-orbital ASAT attack capability. The ASAT development and testing enabled the Soviet Union to have a reliable operational capability from the mid-1970s. The Soviets tested its ASAT capability on over twenty occasions against target satellites in varying orbits and inclinations and operated numerous attack profiles. The intended targets for this ASAT capability were U.S. and NATO satellites. Since the latter stages of the Cold War and indeed since the collapse of the Soviet Union, Russia has shown some interest in an ASAT capability, of most note being the adaptation of the MiG-31 in an ASAT carrier role. Whether this interest will be developed is dependent upon the perceived threat Russia feels in response to the United States' missile defense and ASAT plans.

The United States development of its ASAT capability saw the U.S. Army and Air Force compete for this mission. The initial U.S. ASAT policy utilized the ASAT Program 505, the Nike-Zeus antiballistic missile and was under Army Command. However, when Program 505 was phased out and Program 437, which used the Thor intermediate range ballistic missile, received the ASAT mission, the Air Force took the mission from the Army. This program was later terminated and it was not until the conception of the air launched ASAT that the Air Force continued to hold the ASAT mission. Since the demise of the Soviet Union and the subsequent end of the Cold War, the Army has been at the forefront of ASAT efforts, both in terms of the Ke-ASAT and MIRACL testing. The United States appears to be seriously considering using directed energy for ASAT purposes. The Defense Directive 3100.11 is assessing the world's satellites for their vulnerability to lasers. The safe levels of laser illumination for particular satellites for foreign as well as U.S. domestic satellites is being configured by using computer modeling.

The space-based laser is presently being considered for a weapon system to be actually deployed in space. Interest in using space-based lasers for ballistic missile defense arose with the emergence of two facts. One was that ballistic missiles are relatively fragile and do not resist laser energy and secondly, that chemical lasers could project lethal energy over 3,000 kilometers. The science of utilizing lasers in space has been well documented and the feasibility has been proven. The remaining obstacle in terms of feasibility is the engineering process which needs to undertaken. This process requires funding and a scheduled timetable to achieve this aim.

The companies involved in developing the space-based laser, Lockheed Martin, TRW, and Boeing, appear confident that they would be able to build a working system, and indeed the original schedule to build a demonstration system by 2012, the Integrated Flight Experiment (IFX) appeared to be on track. However, the political climate has altered with a capabilities approach being adopted. This approach has led to a restructuring of the IFX plan with the schedule left with no specific target date being set. The capabilities approach is to identify affordable operational concepts and to focus on early contributions to midcourse and terminal phase intercept systems that are nearer to fruition. This restructuring has left the space-based laser program concentrating on key technologies supporting a future system with no specific focus date, although the restructuring process could lead to redefined experimental space-based laser ahead of the original 2012 schedule date.

The extensive use of space assets during the campaign to liberate Kuwait demonstrated the importance of space systems to modern war. The array of space systems, such as the Global Positioning System, communications satellites and photoreconnaissance satellites enabled battlefield commanders to illuminate the battlefield on such a scale that had not occurred before. The campaign in the Gulf War demonstrated the importance of space assets for the conduct of military operations. These hostilities were the setting for the Revolution in Military Affairs debate, not only in the United States, but for analysts in China and Russia as well. The performance of the United States in particular during the campaign led Russian and Chinese defense analysts to examine whether there was a novel way of conducting war.

The United States, Russian, and Chinese approaches to the Revolution in Military Affairs all identify military space as playing a fundamental part towards achieving a RMA. The components that comprise the RMA are precision strike, information warfare, and dominant maneuver. These components are underpinned by space systems that supply much of the information that is required to illuminate the battlefield in order to take advantage of the aforementioned components. Space is the arena in which the sensors, and transmission of the sensors of the information occurs. The benefit of these transmissions enables the battlefield commander to have an information edge over an adversary. It is not altogether surprising to discover that the United States, along with China and Russia have recognized the importance of space systems to the current RMA.

Military space policy during the Clinton administration received fairly low prioritization. The 1996 National Space Policy did not differ significantly from the previous space policy save for the declassifying of the National Reconnaissance Office. The new directive in 1999 subtly added the proviso "if directed" to the capability to control space, along with the exploration of the concept of space force application. The directive did not specifically announce the development of space control and space force application, merely the research and development of these concepts. The termination of the National Space Council and the dissolving of the deputy undersecretary of defense for space lowered the profile of defense space by merging the previous organizations into nonspecific bureaucracies.

The profile of military space policy under the Bush administration was raised especially with the appointment of Donald Rumsfeld who had chaired the Space Commission, and had spent six months scrutinizing the U.S. space capabilities and requirements. Many organizational changes resulted from the Space Commission and these raise the profile of military space within the policy-making apparatus. The most notable of changes was the appointment of a four-star commander for Air Force Space Command, the creation of a directorate of national security space integration and a national security space architect office.

Conclusion

The issue of military space has been significantly raised in the period following the end of the Cold War. The issue of missile defense has the most significance for the weaponization of space. This issue in particular was politically contentious in both terms of the Clinton administration. The Congressional Missile Defense Acts were met with stringent opposition. This was until President Clinton out of political expediency prior to the presidential elections then accepted the missile defense plan in the form of the "3 plus 3" program. This missile defense architecture utilized an exoatmospheric interceptor that impacts on the threat ballistic missile in space. The lack of success during the initial testing period led the decision of deployment during the Clinton presidency to be deferred to President Bush. The Bush conducted a fundamental review of national missile defense, and removed the "national" from its terminology to announce a widening of missile defense to include allies. The withdrawal from the ABM Treaty demonstrated the new administration's determination to develop a nationwide missile defense.

The issue of ballistic missile defense was one of the key areas that would have led to the weaponization of space. This would have occurred in the United States with the building of a missile defense system that included an exoatmospheric interceptor. This idea gained considerable political support, but this waned considerably in the early 1970s prior to the signing of the ABM Treaty. However, the Soviet Union built their ballistic missile defense system that included the Galosh exoatmospheric interceptor. This system was maintained throughout the Cold War.

Russia under the guise of the Global Protection System devised by Yeltsin showed interest in the early to mid-1990s in cooperating in an international missile defense system. This system would have included space-based weapons. This shows to a certain degree some acceptance of the weaponizing of space. However, the GPS system was no longer under consideration by 1996 and Russian attitudes to missile defense hardened. It is possible that Russia could operationalize the direct-ascent co-orbit antisatellite capability. It is however difficult to put a timeframe on how quickly they could operationalize this. Since the latter stages of the Cold War and indeed since the collapse of the Soviet Union, Russia has shown some interest in an ASAT capability, of most note being the possible adaptation of the MiG-31 in an ASAT carrier role. Whether this interest will be developed upon is dependent upon the perceived threat Russia feels in response to the United States' missile defense and ASAT plans.

China has shown an interest in developing antisatellite weapons, and hence weaponizing space. It is safe to say that the parasitic satellite ASAT is not operational, the microsatellites that are required to carry out this mission are not at the required level of technological maturity. However, it is likely that China has acquired from the former Soviet Union technical know-how to operate the direct ascent method of satellite negation, similar to the co-orbital method of interception. Also, the laser weapon capability of blinding low earth orbit satellites appears to be close to fruition. Certainly there are no technological barriers to China developing such a system. The PRC is certainly seeking to weaponize space. As to what extent this remains a difficult question to answer.

Since the demise of the Soviet Union and susbsequent end of the Cold War, the United States Army has been at the forefront of ASAT efforts, both in terms of the Ke-ASAT and the MIRACL testing. The MIRACL laser testing in 1997 provided valuable information regarding the laser effects on a low earth orbit satellite. This, combined with the Defense Directive that evaluates the world's satellites for the safe levels of illumination, indicates that the United States is seriously considering using lasers as antisatellite weapons.

One effect the RMA could have given the central importance of military space systems to it, is that it could make space a battlefield. As other countries, such as Russia and China see the effects of the RMA and its devastating effects it has on the battlefield, this may lead them to target the space systems themselves. The reliance on space systems to promote the RMA could lead to space becoming a battlefield.

The Space-Based Laser, the IFX plan has however been subject to budgetary considerations. The decision was taken by the Missile Defense Agency to defund the IFX plan. This has meant that there is no longer a scheduled plan to develop a space-based laser. Instead the space-based laser will be funded on a capabilities-based approach, with those missile defense programs closer to operational status being funded. This action will mean a delay in developing a space-based laser and the probable timeframe for an in-orbit space-based laser is now around 2020. Nevertheless, a space-based laser is still under development and is in the United States thinking for weaponizing space. The space-based laser continues to be a politically controversial option for ballistic missile defense.

The issue of missile defense has important implications for the weaponization of space. The ground-based missile defense system under development uses an exoatmospheric hit-to-kill interceptor. This is essentially a space weapon since its intended interception occurs in space. The space-based laser for boost phase missile interception is another space weapon under development. The space-based kinetic kill weapon system, the follow-on to Brilliant Pebbles, is a space weapon that is intended to intercept ballistic missiles in their boost phase. These three weapon systems, all of them space weapons are being developed for the ballistic missile defense role. This is the main rationale for the weaponization of space.

Appendix

The Technological Aspects of Defenses against Antisatellite Weapons (DSATS)

The jamming or blinding of satellite sensors must be carried out by radar, infrared, or visible light sources depending upon the frequency band in which the sensor operates. This method of jamming could be countered by rapidly changing the frequency at which the radar operates, by rapidly shuttering the optical elements of a camera, or other measures.[1] A countermeasure for infrared sensors which are used for launch warning is as follows. Early warning satellites detect the infrared signal from the missile plume as it is picked up above the atmosphere. The missile plume emits over a wide range of frequencies, therefore the frequency which the sensor emits is one which does not penetrate the atmosphere. To jam such a signal would therefore require the jamming source to be attached to the rocket, and hence increase the cost and difficulty.[2] Some protection from directed energy weapons can be achieved by spinning the satellite when it comes under attack, thereby degrading the effects of the energy beam. However, spinning could reduce the performance of many satellites.[3]

One element which is often the focus of attention is the uplink between the ground station and the satellite. These links can be vulnerable to interference. The use of electromagnetic interference, exoatmospheric nuclear detonation, and elimination of communications relay satellites are effective methods of blocking communications. The principle of electromagnetic jamming is saturation of the airways with electronic noise at the same bandwidth that the enemy's communicating with.[4] However, the higher the frequency the more difficult the signal is to jam. As the frequency rises, the beam becomes narrower forcing the jammer to move closer to the receiver or transmitter. A trawler located a few miles offshore from a satellite control facility would be an effective means of jamming communications from several satellites.[5] The advantages of electromagnetic jamming are its potential effectiveness and low cost and risk.

One method of countering jamming is to make the satellites more autonomous by eliminating vulnerable ground stations and providing direct interface with the users.[6] By providing a satellite with more functions it is able to perform onboard specifically with its own command and control and data processing capabilities, the less dependent it becomes upon ground stations which are susceptible to jamming. Also incorporating fault finding software onboard along with redundant, fault-tolerant processors would reduce external control facilities. By adding these capabilities the satellite becomes less dependent on ground control facilities and provides less scope for jamming uplinks.

There is no way to harden a satellite against a nuclear explosion which comes close enough to a satellite. One way to counter this is to distribute the satellite constellation so that one nuclear explosion could destroy no more than one satellite. Satellites are usually thousands of kilometers apart, while the lethal range of a one megaton explosion against a satellite hardened to a feasible level of hardness is under 100 kilometers.[7] However, if the nuclear detonation is further away the electromagnetic pulse effects can be overcome. High-altitude bursts generate an electromagnetic pulse that can couple into unprotected circuits and can cause burnout. Incorporating faraday-cage, filter, surge-arrestor, waveguide cutoff, and fiber-optic technology in the ground site design can provide protection against this threat.[8]

The warning and communication satellites are stationed in high, usually geosynchronous orbits. The targeting of these satellites by ground-launched direct-ascent ASATs would take hours. This time depending upon detection capabilities, would allow the target satellite to take evasive action. The orbital nature of these satellites itself affords them some protection in terms of the time it takes a direct-ascent ASAT to reach the target satellite from the ground. However, a space-based kinetic kill vehicle would have a much shorter time on target and would afford the target satellite much less opportunity to take evasive action.

Low-earth orbit satellites are much easier for ground-based lasers and for direct-ascent ASATs. The time on target for a ground-based ASAT is a matter of minutes and would offer the target satellites little if no opportunity to take evasive action. However, when dealing with ground-based lasers the hardening of satellites would afford the satellites some means of protection.[9] Also as aforementioned, photoreconnaissance satellites could use rapid shuttering to protect the sensitive opto-electric components.

One way of countering the threat from a co-orbital satellite interceptor is to add a maneuver capability to the satellite, especially if the satellite is in a low-medium altitude. The maneuvering of low-orbit satellites greatly complicates the enemy's targeting problems, however if there is insufficient warning time of an impending attack they could still be intercepted. However a disadvantage to the addition of a maneuver capability to a satellite is that payload weight is sacrificed to allow additional fuel to be carried.

One method of providing defenses against possible ASAT usage is to proliferate the number of satellites. That is, provide the enemy with too many targets to intercept and hence cause the enemy to waste resources in countering them. Also, a rapid relaunch capability for critical satellites would also confuse the enemy's targeting plans and could overwhelm its ASAT capabilities, as well as maintaining the crucial satellites in the event of interception.

The Pentagon estimate of the Soviet Union's ASAT interceptor's maximum altitude is 5,000km.[10] The Soviet method of interception to increase its range would have required a larger booster. Alongside this the interceptor would require more electrical power storage, better temperature regulation and better guidance to meet the rigor of higher altitude interception.

The most subtle means of defeating a satellite is by spoofing. Spoofing is either controlling an enemy satellite directly or making the satellite or the ground controller believe that an onboard system need to be controlled when in actuality it does not.[11] If the frequencies, codes, and transmission sequences to control the maneuvered engines are known then an erroneous transmission to fire the engines would cause the satellite to become disoriented, lost, or burn up its fuel unnecessarily. The advantage of spoofing a satellite is that it is possible the enemy would never know what had happened.

The Effects of Nuclear Explosions in Space

There are three characteristics of a nuclear explosion in space which provide it with a possible antisatellite capability: the thermal flash, hard radiation, and electromagnetic pulse (EMP).[12] The effect of the nuclear explosion creates a brief intensive flash of light and heat which can burn through the structure of a satellite at a distance, as well as overheating electronic components and disabling horizon sensors and temperature control systems. The secondary effect is the emitting of radiation in the form of X-rays and neutrons. This radiation penetrates the transistors and creates ions which amass and change the transistor's electrical properties negatively. This could affect the guidance and arming circuits along with the solar cells. The radiation absorbed by the ablative shield on the reentry vehicle could generate enough heat to melt this shield.

The radiation from the explosion is enhanced in space because there is no absorption by the atmosphere.

The electromagnetic pulse effect causes the photons of the initial gamma radiation to exit from the burst at high energies, which collide with the electrons and atoms in the surrounding air and transfer energy to them.[13] The electrons rapidly move away from the center of the burst. This motion, provided that some form of asymmetry exists, is one of the main sources of the electromagnetic pulse. When the electrons move away as a result of the explosion the remaining slower positive ions are left behind. The relative displacement from the negative and positive creates a radial electric field. Under this influence, the large number of electrons are driven back toward the burst point. This produces a second pulse of current, but this is terminated by a recombination of electrons with ions. This large amount of ionized gas (or plasma) oscillates. These oscillations stop in a short time, as the negative particles combine with the positive, but during the oscillation they produce electromagnetic waves in the radio frequency range.

Nuclear bursts at high altitude, around forty kilometers have the effect of jamming satellite communications by absorption or scintillation of the broadcast frequency.[14] This jamming can occur for seconds to hours depending on the transmission frequencies. The higher the frequency the shorter the interruption. A single detonation can have enormous effects. For example, a one-megaton detonation at 100km above the central United States could block UHF communications for around thirty minutes over the entire country.[15]

Notes

1. Michael M. May, "Safeguarding Our Space Assets," in William Perry, et al, *Seeking Stability in Space: Antisatellite Weapons and the Evolving Space Regime,* (Aspen: Univeristy Press of America, 1984), 72.
2. May, 72-74.
3. Robert. B. Giffen, "Space System Survivability," in Uri Ra'anan and Robert L., Pfaltzgraff, *International Security Dimensions of Space* (Hamden, CT: Archon Books, 1984), 88
4. Giffen, 84.
5. Giffen, 85.
6. Giffen, 91.
7. Giffen, 77.
8. Giffen, 88.
9. Giffen, 81.

10. Donald L. Hafner, "Negotiating Restraints on Antisatellite Weapons: Options and Impact," in *Seeking Stability in Space: Antisatellite Weapons and the Evolving Space Regime,* (Hamden, CT: Archon Books, 1984), 98.

11. Robert B. Giffen, "Space System Survivability," in Uri Ra'anan and Robert L. Pfaltzgraff, *International Security Dimensions of Space*, 81.

12. Curtis Peebles, *Battle for Space* (Dorset: Blandford Press, 1983), 80.

13. Samuel Glasstone, ed., *The Effects of Nuclear Weapons* (U.S. Department of Defense: U.S. Government Printing Office, 1962), 503.

14. Robert B. Giffen, "Space System Survivability," in Uri Ra'anan and Robert L. Pfaltzgraff eds., *International Security Dimensions of Space*, 85.

15. Giffen, 85.

Bibliography

Government Documents

Augenstein, Bruno W., *Evolution of the U.S. Military Space Program 1945-60: Some Key Events in Study, Planning, and Program Management* (Santa Monica, CA: RAND, 1982).

Buchan, Glenn C., *The Impact of the Revolution in Military Affairs on Developing States" Military Capability* (Santa Monica, CA: RAND, July 1995).

———. *One-and-a-Half Cheers for the Revolution in Military Affairs* (Santa Monica, CA: RAND, 1998).

Bush, George Sr, President, *Presidential State of the Union Address*, January 29, 1991.

Carter, Ashton B., *Directed Energy Missile Defense in Space*, Background Paper (Washington, D.C.: U.S. Congress, Office of Technolgoy Assessment, OTA-BP-ISC-26, April 1984).

China's National Defense (Beijing, Information Office of the State Council of The People's Republic of China: 1998).

Cohen, William S., Secretary of Defense, *Annual Report to the President and the Congress* (Washington, D.C.: U.S. GPO, 1999).

Commission to Assess U.S. National Security Space Management and Organization (Washington, D.C.: U.S. GPO, January 11, 2001).

Defend America Act of 1996 (Washington, D.C.: U.S. GPO, March 1996).

Department of Defense, *Department of Defense Space Policy* (Washington, D.C.: U.S. GPO, 1987).

———. *Soviet Military Power 1989* (Washington, D.C.: U.S. GPO, 1989).

———. *Conduct of the Persian Gulf War, Final Report to Congress* (Washington, D.C.: U.S. GPO, 1992).

Department of Defense Directive, *Space Policy*, Number 3100.10, July 9, 1999.

Defense Intelligence Agency, *Soviet Military Space Doctrine* (Washington, D.C.: U.S. GPO, 1984).

"Directed and Kinetic Energy Systems Technology," *DTIC Report*, March 1999, Military Critical Technologies, http://www.iac.dtic.mil/mctl/data/sec04.pdf.

"Dragons in Orbit? Analyzing the Chinese Approach to Space," *National Defense University*, Washington, D.C., August, 2001 http://www.ndu.edu/inss/China_Center/paper10.htm.

Estes, Howell M., *U.S. Space Command Long Range Plan* (Colorado Springs: Headquarters, U.S. Command, 1998).

Falto-Heck, Ivette, Colonel USAF, System Program Director, Space-Based Laser Project Office, Space and Missiles Center, Air Force Space Command, 'space-Based Laser (SBL) Requirements for Experimental Missile Defense System," Presentation at *SMi Military Battle Space Conference*, May 29, 2002.

Final Report of the Select Committee on U.S. National Security and Military/Commercial Concerns with the People's Republic of China (Washington, D.C.: U.S. GPO, 1999).

Glasstone, Samuel, ed., *The Effects of Nuclear Weapons* (U.S. Department of Defense: U.S. Government Printing Office, 1962).

Gonzales, Daniel, *The Changing Role of the U.S. Military in Space* (Washington D.C.: RAND, 1999).

Gonzales, David, Mel Eisman, Calvin Shipbaugh, Timothy Bonds and Anh Tuan Le, *Proceedings of the RAND Project Air Force Workshop on Transatmospheric Vehicles*, Rand Report MR-890-AF (Santa Monica, CA: RAND, 1997).

Hadley, Stephen J., and Henry Cooper, *Briefing on the Refocused Strategic Defense Initiative*, February 12, 1991.

Hildreth, Stephen A., *The Strategic Defense Initiative: Issues for Phase I Deployment*, Congressional Research Service Issue Brief (Washington, D.C.: CRS, 1990).

Johnson, Dana J., *Space: Emerging Options for National Power* (Santa Monica, CA: RAND, 1998).

Joint Vision 2010, Office of the Chairman of the Joint Chiefs of Staff, Joint Staff (Washington, D.C.: Pentagon, April 1996).

National Missile Defense Act of 1999, 106th Congress, 1st Session, House Resolution 4 (Washington, D.C.: U.S. GPO, 1999).

Office of the Press Secretary to the White House, *ABM Treaty Factsheet*, December 13, 2001.

Office of Public Affairs, *Lab Evaluate Satellites*, United States Air Force Research Laboratory, July 6, 2000 http://www.de.afrl.af.mil/News2000/00-50.html.

Office of Technology Assessment, *Antisatellite Weapons, Countermeasures, and Arms Control: Summary* (Washington, D.C.: U.S. GPO, 1985).

Berman, Robert P. and John C. Baker, *Soviet Strategic Forces: Requirements and Responses* (Washington, D.C.: Brookings Institute, 1982).

Binnenendijk, Hans and Ronald N. Montaperto, *Strategic Trends in China*, Institute for National Strategic Studies (Washington, D.C.: National Defense University, 1998).

Bonds, Ray, ed., *Soviet War Power* (London: Corgi, 1982).

Bulkeley, Rip and Graham Spinardi, *Space Weapons: Deterrence or Delusion?* (Cambridge: Policy Press, 1986).

Burrows, William E., *Deep Black: Space Espionage and National Security* (New York: Berkley Books, 1986).

Burrows, William E., *This New Ocean* (New York: Random House, 1998).

Byerly, Radford, ed., *Space Policy Reconsidered* (Boulder, CO: Westview Press, 1989).

Callaham, M. and K. Tsipis, *High Energy Laser Weapons: A Technical Assessment* (Cambridge, MA: Massachusetts Institute of Technology, 1980).

Canan, James, *War in Space* (New York: Harper and Row Publishers, 1982).

Codevilla, Angelo, *While Others Build* (New York: Free Press, 1988).

Collins, John M., *Military Space Forces: The Next Fifty Years* (McLean, VA: Pergmon Brassey's, 1989).

Day, Dwayne, John B. Logsdon, and Brian Latell, eds., *Eye In The Sky: The Story of the Corona Spy Satellites* (London: Smithsonian Institute Press, 1998).

Davis, William, Jr, *Asymmetries in U.S. and Soviet Strategic Defense Programs* (Pergammon Brasseys: Washington, D.C., 1986).

Deane, Michael J., *The Role of Strategic Defense in Soviet Strategy* (Coral Gable: University of Miami Press, 1980).

DeBlois, Bruce, *Beyond The Paths of Heaven: The Emergence of Space Power Thought* (Alabama: Maxwell Air Force Base, 1999).

Dietrick, Kevin M., Lt. Col., *Whence the Army's Role in Space* (Carlisle Barracks, PA: U.S. Army War College, 2001).

Duffner, Robert W., *Airborne Laser: Bullets of Light* (New York: Plenum Press, 1997).

Durch, William J., *National Interest and the Military Use of Space* (Cambridge, MA: Ballinger Publishing, 1984).

Dutton, Lyn, *Military Space* (London, UK,: Brasseys, 1991).

Emme, Eugene M., ed., *The Impact of Air Power: National Security and World Politics* (Princeton, NJ: D. Van Nostrand, 1959).

Fast Scott, Harriet, 3rd ed. V.D. Sokolovsky *Soviet Military Strategy* (New York: Crane, Russak, 1975).

Frances Fitzgerald, *Way Out There in the Blue: Reagan, Star Wars and the End of the Cold War* (New York: Simon and Schuster, 2000).

Garis, David de, *Military Space* (London: Brassey's Air Power Series: 1991).

Gibson, Robert D., Lt. Col. (USAF) *Space Power, The Revolution in Military Affairs* (Carlisle Barracks, PA: U.S. Army War College, 2001).

Gill, Bates and Lonnie Henley, *China and the Revolution in Military Affairs* (Carlisle Barracks, PA: U.S. Army War College: 1996).

Goldman, Nathan C., *Space Policy An Introduction* (Iowa: Iowa State University Press, 1992).

———. *American Space Law* (Iowa: Iowa State University Press, 1988).

Gongora, Thierry and Harald von Riekhoff, *Toward A Revolution in Military Affairs Defense and Security at the Dawn of the 21st Century* (Westport, CT: Greenwood Publications, 2000).

Gordon, Michael R., and General Bernard R. Trainor, *The General's War* (New York: Little, Brown and Co., 1994).

Graham, Daniel O., *The Non-Nuclear Defense of Cities: The High Frontier Space-Based Defense against ICBM Attack* (Cambridge, MA: Abt Books, 1983).

Gray, Colin S., *American Military Space Policy: Information Systems, Weapons Systems and Arms Control* (Cambridge, MA: Abt Book, 1983).

Grossman, Karl, *Weapons in Space* (New York: Seven Stories Press, 2001).

Gurtov, Mel and Byong-Moo Hwang, *China's Security: The New Roles of the Military* (London: Lynne Rienner Publishing, 1998).

Hall, R. Cargill, and Jacob Neufeld, *The U.S. Air Force in Space 1945 to the Twenty-First Century* (Washington, D.C.: USAF History and Museums Program, 1998).

Harvey, Brian, *The New Russian Space Program: From Competiton to Collaboration* (Chichester: Praxis Publishing, 1996).

———. *The Chinese Space Program: From Conception to Future Capabilities* (Chichester: Praxis Publishing, 1998).

———. *Russia in Space: The Failed Frontier?* (Chichester: Praxis Publishing, 2001).

Hawkins, Charles F., "The Four Futures: Competing Schools of Thought Inside the PLA," *Taiwan Security Research*, March 2000 http://taiwansecurity.org/IS/IS-0300-Hawkins.htm.

Hays, Peter, *Struggling Towards Space Doctrine: U.S. Military Space Plans, Program, and Perspectives During the Cold War*, Ph.D. Thesis, Fletcher School of Law and Diplomacy, 1994.

———. *Space Power Interests* (Boulder, CO: Westview Press, 1996).

Holmes, Kim R. and Baker Spring, eds., *SDI at the Turning Point: Readying Strategic Defenses for the 1990s and Beyond* (Washington D.C.: The Heritage Foundation, 1990).

Hunter, R. C., *United States Antisatellite Policy for a Multipolar World* (Maxwell Air Force Base, AL: Air University Press, 1995).

International Institute for Strategic Studies, *The Military Balance 1998-99* (Oxford: Oxford University Press, 1999).

James, P. N., *The Soviet Conquest in Space* (New York: Arlington House, 1974).

Jasani, Bhupendra, *Outer Space—Battlefield of the Future?* (Solna, Sweden: SIPRI, 1978).

———. *The Militarization of Space: An Arms Control Dilemma* (Basingstoke: Macmillan Press, 1988).

Ji, You, *The Armed Forces of China* (New York: I. B. Taurus Publishers, 1999).

Johnson, Dana L., *The Evolution of Military Space Doctrine: Precedents, Prospects, and Challenges*, Ph.D Thesis, University of Southern California, December 1987.

Johnson, Nicholas J., *Soviet Military Strategy in Space* (London: Jane's Publishing Company, 1987).

Johnson-Freese, Joan, *The Chinese Space Program: A Mystery Within a Maze* (Malabar, FL: Krieger Publishing: 1998).

Johnson-Freese, Joan and Roger Handberg, *Space the Dormant Frontier: Changing the Paradigm for the 21st Century* (Westport, CT: Praeger Publishing, 1997).

Johnson-Freese, Joan, *The Viability of U.S. Antisatellite Policy: Moving Toward Space Control*, Institute for National Security Studies Occasional Paper 30, 2000.

Klass, Philip J., *Secret Sentries in Space* (New York: Random House, 1971).

Kirby, Stephen et al, *The Militarization of Space* (Sussex: Lynne Rienner, 1987).

Labrie, Roger P., ed., *SALT Handbook: Key Documents and Issues* (Washington, D.C., American Enterprise Institute: 1979).

Lakoff, Sanford and Randy Willoughby, *Strategic Defense and the Western Alliance* (Massachusetts, Lexington Books: 1987).

Lambakis, Steven, *On the Edge of Earth: The Future of American Space Power* (Lexington: University Press of Kentucky, 2001).

Lee, Christopher, *War in Space* (London: Hamish Hamilton, 1986).

Lilley, James R. and Daria Shambaugh, eds., *China's Military Faces The Future* (Washington, D.C.: American Enterprise Institute for Public Policy Research, 1999).

Lindsey, James M. and Michael E. O"Hanlon, *Defending America: The Case for Limited National Missile Defense* (Washington, D.C.: Brookings Institute Press, 2001).

Long, Franklin A., Donald Hafner, and Jeffrey Boutwell eds., *Weapons in Space* (New York: Norton, 1986).

Lupton, David E., *On Space Warfare: A Space Power Doctrine* (Alabama, Maxwell Air Force Base: Air University Press, 1988).

Luongo, Kenneth and W. Thomas Wander, *The Search for Security in Space* (Ithaca: Cornell University Press, 1989).

Manno, Jack, *Arming the Heavens. The Hidden Military Agenda for Space* (New York: Dodd, Mead and Company, 1984).

Mathers, Jennifer G., *The Russian Nuclear Shield from Stalin to Yeltsin* (London: Macmillan Press, 2000).

McDaniel, Ruth Currie, *The U.S. Army Strategic Defense Command: Its History and Role in the Strategic Defense Initiative,* 2nd ed. (Huntsville, AL: U.S. Army Strategic Defense Command, 1987).

McDonald, Robert A., *Corona Between The Sun and The Earth The First NRO Reconnaissance Eye in Space* (Maryland: American Society for Photogrammetry, 1997).

McDougall, Walter A., *The Heavens and the Earth: A Political History of the Space Age* (New York: Basic Books, 1985).

McLean, Alasdair, *The Military Use of Space* (Aberdeen: Center for Defense Studies, 1991).

———. *West European Military Space Policy* (Brookfield, VT: Vermont University Press, 1992).

Milton, A. F., ed., *Making Space Defense Work* (London: Pergammon Brasseys, 1989).

Nguyen, Hung P., *Submarine Detection from Space—A Study of Russian Space Capabilities* (London: Shrewsburys Airlife Publishing, 1994).

O'Hanlon, Michael, *Technological Change and the Future of Warfare* (Washington, D.C:, Brookings Institution Press, 2000).

Papp, Daniel S. and John R. McIntyre, *International Space Policy: Legal, Economic, and Strategic Options for the Twentieth Century and Beyond* (New York: Quorum Books, 1987).

Parrott, Bruce, *The Soviet Union and Ballistic Missile Defense* (Boulder, CO: Westview Press, 1987).

Payne, Keith B., *Laser Weapons in Space: Policy and Doctrine* (Boulder, CO: Westview Press, 1983).

———. *Strategic Defense: "Star Wars" in Perspective* (Lanham, MD: The Hamilton Press, 1986).

Peebles, Curtis, *Battle for Space* (Dorset: Blandford Press, 1983).

———. *Guardians: Strategic Reconnaissance Satellites* (California: Presido Press, 1987).

Perry, William J., Brent Scowcroft, Joseph S. Nye and James A. Schear, *Seeking Stability in Space: Antisatellite Weapons and the Evolving Space Regime* (Aspen: University Press of America, 1987).

Pillsbury, Michael, *Chinese Views of Future Warfare*, The Institute for National Strategic Studies (Washington, D.C.: National Defense University, 1998).

Pillsbury, Michael, *China Debates The Future Security Enviroment* (Washington D.C.: National Defense University, 2000).

Ra"anan, Uri and Robert L. Pfaltzgraff, Jr., eds., *International Security Dimensions of Space* (Hamden, CT: Archon Books, 1984).

Reynolds, G. Y., *Outer Space, Problems of Law and Policy* (Boulder, CO: Westview Press, 1989).

Rhea, John, ed., *Roads To Space: An Oral History of the Soviet Space Program* (New York: McGraw-Hill, 1995).

Salkeld, Robert, *War And Space* (New Jersey: Prentice Hall, 1970).

Schaerf, Carlo, Giuseppe Longo and David Carlton, *Space and Nuclear Weaponry in the 1990s* (London: Macmillan, 1992).

Sellers, J. J., *Understanding Space an Introduction to Astronautics* (Portland, OR: McGraw-Hill, 1993).

Shaffer, Stephen M. and Lis Robock Shaffer, *The Politics of International Cooperation: A Comparison of U.S. Experience in Space and in Security*, Volume 17 Book 4 (Monograph Series in World Affairs: University of Denver, 1980).

SIPRI Yearbook 2002 (Stockholm International Peace Research Institute: Oxford University Press, 2002).

Spacy, William L., Major USAF, "Does the United States Need Space-Based Weapons?" *College of Aerospace Doctrine, Research and Education* (Maxwell Air Force Base: Air University Press, 1999).

Stares, Paul B., *The Militarization of Space* (Ithaca: Cornell University Press, 1985).

Stares, Paul B. and Michael Schwartz, *The Exploitation of Space* (London: Butterworths, 1985).

Stares, Paul B., *United States and Soviet Space Programs: A Comparative Assessment* (Ithaca: Cornell University Press, 1989).

Stevens, Sayre, *Ballistic Missile Defense* (Washington, D.C.: Brookings Institution, 1984).

Stine, Harry G., *Confrontation in Space* (New Jersey: Prentice Hall, 1981).

Stokes, Mark A., *China's Strategic Modernisation: Implications for the United States* (Carlisle, PA: U.S. Army War College: Strategic Studies Institute, September 1999).

Sturm, Thomas A., *The USAF Scientific Advisory Board: Its First Twenty Years, 1944-1964* (Washington, D.C.: Government Printing Office, 1986).

Summary of the Center for Security Policy's High-level Roundtable Discussion of: "Space Power: What is at Stake, What will it Take?" 11 December 2000 902 Hart Senate Office Building, Washington, D.C., http://www.centerforsecuritypolicy.org/papers/2001/01-P04at.shtml.

Tanks, David R., *National Missile Defense: Policy Issues and Technological Capabilities* (Washington, D.C.: Institute for Foreign Policy Analysis, 2000).

Thomas, Timothy L., *Behind the Great Firewall of China: A Look at RMA/IW Theory from 1996-1998* (Fort Leavenworth, KS: Foreign Military Studies Office, 1998).

Thompson, Col. David J. and Lt. Col. William R. Morris, *China in Space Civilian and Military Developments* Air War College Maxwell Paper No. 24 (Alabama: Maxwell Air Force Base, 2001).

Weinberger, Caspar W., *Fighting for Peace: Seven Critical Years in the Pentagon* (New York: Warner Books, 1990).

Werrell, Kenneth P., *Hitting a Bullet with a Bullet: A History of Ballistic Missile Defense* (United States Airpower Research Institute, 2000).

Yanarella, Ernest J., *The Missile Defense Controversy: Strategy, Technology and Politics, 1955-1972* (Lexington: University Press of Kentucky, 1977).

York, Herbert, *Race to Oblivion* (New York: Simon and Schuster, 1970).

Yost, David, *Soviet Ballistic Missile Defense and the Western Alliance* (Cambridge and London: Harvard University Press, 1988).

Ziegler, David W., *Safe Heavens: Military Strategy and Space Sanctuary Thought* (Maxwell Air Force Base, AL: Air University Press, 1998).

Articles

"Air Force Lab Evaluating Satellites Vulnerability t Lasers," *Space Daily*, July 12, 2000 http://www.spacedaily.com/news/laser-00i.html.

"Alpha Missile Defense Laser Is Fired for First Time," *Aviation Week and Space Technology*, April 19, 1989.

Anselmo, Joseph C., "New Funding Spurs Laser Efforts," *Aviation Week and Space Technology*, October 14, 1996.

———. "Missile Test Extends North Korea's Reach," *Aviation Week and Space Technology*, September 7, 1998.

Anson, Peter, Rr. Adm. RN, and Dennis Cummings, Cpt, RAF, "The First Space War: The Contribution of Satellites to the Gulf War," *RUSI Journal*, Winter 1991.

Arquilla, John, "The Strategic Implications of Information Dominance," *Strategic Review*, Summer 1994.

"Asia: A Divine Lift-off for China," *The Economist*, November 27, 1999.

Baozhu, Guo, "Development of China's Space Industry at the Turn of the New Century," *Aerospace China* (English edition), Summer 2001.

Bender, Bryan, "U.S. Blueprint for Future Weapons Systems Is Outlined," *Jane's Defense Weekly*, May 26, 1999.

Bergman, Kenneth R., "Space and the Revolution in Military Affairs," *Marine Corps Gazette*, May 1995.

Bian, Wu, "Space Industry Promotes Modernisation," *Beijing Review*, January 6-12, 1997.

Bianji, "China Launching Its First Navigation and Position Satellite," *Aerospace China* (English edition), Winter 2000.

Biddle, Stephen, "Victory Misunderstood: What the Gulf War Tells Us about the Future of Conflict," *International Security*, Fall 1997.

Blaker, James R., "Understanding the Revolution in Military Affairs: A Guide to America's 21st Century Defense," *Progressive Policy Intsitute*, January 1997.

Blaker, James R., "A Vanguard Force: Accelerating the American Revolution in Military Affairs," *Progressive Policy Institute*, Defense Working Group Policy Brief, November 1997.

Bruger, Steven, Lt. Col., USAF, "Not Ready for the First Space War, What about the Second?" *Naval War College Review*, Winter 1995.

Caceres, Marco A., "Satcom Market Buffeted by Economic Uncertainties," *Aviation Week and Space Technology*, January 11, 1999.

Changchui, He, "The Development of Remote-sensing in China," *Space Policy*, February 1989.

Chen, Yanping, "China's Space Policy—A Historical Review," *Space Policy*, May 1991.

———. "China's Space Program," *Ad Astra*, September 1991.

"China Pursues Western Technology to Augment National Space Programs," *Aviation Week and Space Technology*, March 18, 1985.

"Chinese Satellite Splashes into the Pacific," *Aerospace America*, December 1993.

"China Aerospace Declaration Facing New Century," *Aerospace China* (English edition), Winter 2000.

"China Eyes New Spaceport and Bigger Rockets," *Space Daily*, February 6, 2001, http://www.spacedaily.com/news/china-01t.html.

"China launches maiden navigation positioning satellite," *Space Daily*, October 31, 2001, htttp://www.spacedaily.com/news/001031025809.g4qwq3ac.html.

"China launches second navigation position satellite," *Space Daily*, December 31, 2000, http://www.spacedaily.com/news/001221032826.9hwr6xru.html.

"China pledges to build satellite network, put man in space," *Space Daily*, November 22, 2000, http://www.spacedaily.com/news/001122095908.q7jwj45w.html.

"Chinese experts advocate new rocket launching center," *Space Daily*, February 6, 2001, http://www.spacedaily.com/news/0120206135706.n1zkucw1.html.

Chu, Shulong, "The PRC Girds for Limited, High Tech War," *Orbis*, Spring 1994.

Clark, Phillip, "Chinese Designs on the Race for Space," *Jane's Intelligence Review*, April 1997.

Cooper, Henry F., "Antisatellite Systems and Arms Control Lessons From The Past," *Strategic Review*, Spring 1981.

———. "Unsteady Evolution of the Emerging Consensus on SDI," *Comparative Strategy*, January-March, Volume 12 Number 1, 1993.

Cooper, Pat, "U.S. Political Battles Threaten Antisatellite Project," *Space News*, June 17-24, 1996.

Cosyn, Phillipe, "China Plans Rapid-Response, Mobile Rocket, Nanosatellite Next Year," *Space Daily*, May 1, 2001, http://www.spacedaily.com/news/china-01zc.html.

Covault, Craig, "Austere Chinese Space Program Keyed Toward Future Buildup," *Aviation Week and Space Technology*, July 8, 1985.

———. "Chinese Facility Combines Capabilities to Produce Long March Boosters, ICBMs," *Aviation Week and Space Technology*, July 27, 1987.

———. "Chinese Space Program Sets Aggressive Pace," *Aviation Week and Space Technology*, October 5, 1992.

———. "Russian Military Space Program Maintains Aggressive Pace," *Aviation Week and Space Technology*, May 3, 1993.

———. "China Seeks Cooperation, Airs New Space Strategy," *Aviation Week and Space Technology*, October 14, 1996.

———. "Commercial Proton, Soyuz Launch Surge Readied," *Aviation Week and Space Technology*, February 8, 1999.

———. "Promise and Peril Mark Russian Launch Surge," *Aviation Week and Space Technology*, July 13, 1998.

———. "Military Space Dominates Air Strikes," *Aviation Week and Space Technology*, March 29, 1999.

———. "Recon, GPS Operations Critical to NATO Strikes," *Aviation Week and Space Technology*, April 26, 1999.

———. "Manned Program Advances Chinese Space Technology," *Aviation Week and Space Technology*, November 29, 1999.

Covault, Craig, "Naval Space Ops Crucial to Afghan War," *Aviation Week and Space Technology*, April 8, 2002.

Crouch, J. D. "SDI and Securing Western Freedom," *Laissez Faire*, Volume 1 Number 4, Summer 1992.

Daggert, Stephen and Robert D. English, "Assessing Soviet Strategic Defense," *Foreign Policy*, Spring 1988.

Day, Dwayne A., "Capturing the High Ground The U.S. Military in Space 1987-1995 Part I," *Countdown*, January/February 1995.

———. "Capturing the High Ground The U.S. Military in Space 1987-1995 Part II," *Countdown*, May/June 1995.

DeBiaso, P. A., 'space-Based Defense," *Comparative Strategy*, January-March, volume 12 number 1, 1993.

Dengrui, Liu, "China's Space Industry Forging Ahead," *China Today*, September 1996.

"Defense Department Details Chinese Military Space Capabilities and Plans," *Space Daily*, June 28, 2000, htttp://www.spacedaily.com/news/china-milspace-00a.html.

Dole, Senator Robert, "U.S.-Russia Should Build Joint Missile Defense," *USA Today*, May 9, 1995.

Dooling, Dave, "Space Sentries," *IEEE Spectrum*, September 1997.

Dornheim, Michael A., "Missile Destroyed in First SDI Test at High Energy Laser Facility," *Aviation Week and Space Technology*, September 23, 1985.

———. "Laser Engages Satellite, with Questionable Results," *Aviation Week and Space Technology*, October 27, 1997.

———. "Pentagon Mulls Space Laser Test," *Aviation Week and Space Technology*, March 23, 1998.

Durch, William J., "The Future of the ABM Treaty," *Adelphi Papers*, Summer 1987.

"Excerpts from Speech by Gorbachev about Iceland Meeting," *New York Times*, October 15, 1986.

Feigenbaum, Evan A., "Who's behind China's High Technology "Revolution"?" *International Security*, Summer 1999.

Filho, Jose Monserrat, "Brazilian-Chinese Space Cooperation: An Analysis," *Space Policy*, May 1997.

Fisher, Richard D., "China Increases Its Missile Forces While Opposing U.S. Missile Defense," *The Heritage Foundation*, April 7, 1999.

———. "How America's Friends Are Building China's Military Power," *The Heritage Foundation*, November 5, 1997.

Fitzgerald, Mary C., "The Soviet Military and the New "Technological Operation" in the Gulf," *Naval War College Review*, Autumn, 1991.
———. "The Russian Military Strategy For "Sixth Generation" Warfare," *Orbis*, 1994.
France, Martin E. B., "Back to the Future: Space Power Theory and A. T. Mahan," *Space Policy*, February 2000
Freedman, Lawrence, "The Soviet Union and "Anti-Space Defense"," *Survival*, January/February 1977.
———. "Sanctuary or Combat Zone? Military Space in the 21st Century," Air Power and Space-Future Perspectives Conference, September 12-13 1996.
———. "Britain and the Revolution in Military Affairs," *Revolution in Military Affairs? Challenges to Governments and Industry in the Information Age*, Conference Paper, Royal Institute for International Affairs, May 21 and 22, 1997.
———. "The Revolution in Strategic Affairs," *Adelphi Papers*, Number 318, International Institute for Strategic Studies, 1998.
Friedberg, Aaron L., "Arming China Against Ourselves," *Commentary*, July-August 1999.
Fulghum, David A., "Laser Offers Defense against Satellites," *Aviation Week and Space Technology*, October 7, 1996.
———. "North Korea Space Attempt Verified," *Aviation Week and Space Technology*, September 21, 1998.
Gaffney, Frank J., Jr., "Wake-Up Call on Space," *CNSNews*, January 9, 2001 http://usconservatives.about.com/blc0109space.htm
Gartoff, Raymond, "Banning the Bomb in Outerspace," *International Security*, Winter 1980/81.
Garwin, Richard L., "National Security Space Policy," *International Security*, Spring 1987.
Garwin, Richard L., Kurt Gottfried and Donald L. Hafner, "Antisatellite Weapons," *Scientific American*, June 1984.
Gelb, Leslie, "Reagan Reported to Stay Insistent on "Star Wars" Test," *New York Times*, October 15, 1986.
Gerth, Jeff and Raymond Bonner, "Companies Are Investigated for Aid to China on Rockets," *New York Times*, April 4, 1998.
Gertz, Bill, "Star Wars Backer Hail Defense Project with Russia," *Washington Post*, February 21, 1992.
———. "Clinton, Yeltsin Agree on Missiles," *Washington Times*, May 11, 1995.
Gilks, Anne, "China's Space Policy: Review and Prospects," *Space Policy*, August 1997.

Gill, Bates and Evan S. Medeiros, "Foreign and Domestic Influences on China's Arms Control and Nonproliferation Policies," *The China Quarterly*, March 2000.

Graham, Bradley, "Congress to Push for a National Missile Defense," *Washington Post*, September 5, 1995.

Grahame, David, "A Question of Intent: Missile Defense and the Weaponization of Space," *British American Security Information Council*, May 1, 2002, http://www.basicint.org/NMDSpace.htm.

Grace, Jim, "Global Positioning Guidance Boosts Projectile Proficiency," *Signal*, December 1999.

Graham, D. R., "Missile Defense Capability," *Comparative Strategy*, January-February, Volume 12 Number 1, 1993.

Gray, Colin S., "The Military Uses of Space: Space is not a Sanctuary," *Survival*, Volume XXV Number 5, September/October 1983.

———. 'space Power Survivability," *Airpower Journal*, Winter 1993.

———. "A Contested Vision: The RMA Debate Today," *Revolution in Military Affairs? Challenges to Governments and Industry in the Information Age*, Conference Paper, Royal Institute for International Affairs, May 21-22 1997.

Gray, Colin S. and John B. Sheldon. "Space Power and the Revolution in Military Affairs: A Glass Half Full?" *Air Power Journal*, Fall 1999.

Grou, Lester W. and Timothy L. Thomas, "A Russian View of Future War: Theory and Direction," *Journal of Slavic Military Studies*, September 1996.

Guoxiand, Wu, "China's Space Communications Goals," *Space Policy*, February 1988.

Hadley, Stephen, "Global Protection System: Concept and Process," *Comparative Strategy*, Volume 12 Number 1, January-February, 1993.

Hafner, Donald L., "Averting A Brobdingnagian Skeet Shoot," *International Security*, Winter 1980/81.

Hamon, Dale R., Commander, USN, Ret. and Lt. Col., ret. G. Walter, USAF, 'space and Power Projection," *Military Review*, November 1994.

Hansen, Richard Earl, Lt. Col., "Freedom of Passage on the High Seas of Space," *Strategic Review*, Fall, 1977.

Hewish, Mark, "The Sensor of Choice: Synthetic Aperture Radar," *Jane's International Defense Review*, May 1997.

Hickman, John and Evertt Dolman, "Resurrecting the Space Age: A State Centerd Commentary on the Outer Space Regime," *Comparative Strategy*, January-March 2002.

Hitchens, Theresa, "Rushing to Weaponise the Final Frontier," *Arms Control Today*, September 2001, http://www.armscontrol.org/act/2001_09/hitchenssept01.asp.
———. "Weapons in Space: Silver Bullet or Russian Roulette?" *Center for Defense Information*, April 18, 2002 http://www.cdi.org/missile-defense/spaceweapons.cfm.
Ho, Cheng, "China Eyes Antisatellite System," *Space Daily*, July 8, 2000, http://www.spacedaily.com/news/china-01c.html.
———. "First Chinese Navsat in Operation," *Space Daily*, November 22, 2000, http://www.spacedaily.com/news/china-00zzr.html.
Hughes, James H., "Chinese Space Power," *Journal of Social, Political and Economic Studies*, Summer 2000.
———. "A Vision for Space," *Journal of Social, Political and Economic Studies*, Spring 2000.
Jan-Ping, Wu, Col. (ROC), "The People's Liberation Army in the 21st Century," *RUSI Journal*, June 2000.
Jasani, Bhupendra, "U.S. National Missile Defense and International Security: Blessing or Blight?" *Space Policy*, November 2001.
Ji, You, "The Revolution in Military Affairs and the Evolution of China's Strategic Thinking," *Contemporary Southeast Asia*, December 1999.
Jiasheng, Wang, "China's Communications Satellites," *Aerospace China*, (English Edition), Autumn 2001.
Jiyuan, Liu, "Space For Development Launch Services in China's Space Program," *Harvard International Review*, Summer 1994.
Johnson, Alastair Iain, "Prospects for Chinese Nuclear Force Modernisation: Limited Deterrence Versus Multilateral Arms Control," *China Quarterly*, 1996.
Karniol, Robert, "Power to the People," *Jane's Defense Weekly*, July 12, 2000.
Kipp, Jacob W., "Russian Military Forecasting and the Revolution in Military Affairs: A Case of the Oracle of Delphi or Cassandra," *Journal of Slavic Military Studies*, March 1996.
———. "Confronting the RMA in Russia," *Military Review*, May-June, 1997.
———. "The Revolution in Military Affairs and its Interpreters: Implications for National and International Security Policy," *Foreign Military Studies Office*, September 1995.
Kitfield, Kames, "The Permanent Frontier," *The National Journal*, March 17, 2001,
http://www.globalsecurity.org/org/news/2001/010317-nj.html.
Krepon, Michael, 'spying From Space," *Foreign Policy*, Summer 1989.

Lambakis, Steven, "Space Control in Desert Storm and Beyond," *Orbis*, Summer 1995.

Langereux, Pierre and Christian Lardier, "Launch Setbacks Fail to Dent China's Space Ambitions," *Interavia Business and Technology*, December 1996.

Lavitt, Michael O., "China to Expand Satellite Line," *Aviation Week and Space Technology*, March 29, 1993.

Lawrence, Susan V, "Celestial Reach," *Far Eastern Economic Review*, December 2, 1999.

Libicki, Martin C., "Information Warfare, Information Peace," *Journal of International Affairs*, Volume 51 Number 2, Spring 1998.

Logsdon, John M., "Just Say Wait to Space Power," *Issues in Science and Technology*, Spring 2001.

Long, Wei, "Ambitious Space Effort Challenges China in Next Five Years," *Space Daily*, September 18, 2001, http://www.spacedaily.com/news/china-01zw2.html.

———. "China Builds Advanced Spacecraft Tracking and Command Network," *Space Daily*, May 29, 2000, http://www.spacedaily.com/news/china-00za.html.

———. "China's Plans Beyond Shenzhou," *Space Daily*, November 30, 2000, http://www.spacedaily.com/news-00zzs.html.

———. "China's Space Program Faces Serious Brain Drain," *Space Daily*, March 13, 2000, http://www.spacedaily.com/news/china-00k.html.

———. "Chinese Government Backs Commercial Space Push," *Space Daily*, October 13, 2000, http://www.spacedaily.com/news/china-00zzh.html.

———. "China, Brazil Continue Remote-sensing," *Space Daily*, September 28, 2000 http://www.spacedaily.com/news/china-00zzf.html.

Luttwak, Edward N., "A Post-Heroic Military Policy," *Foreign Affairs*, July/August 1996

Mahnken, Thomas G., "Why Third World Space Systems Matter," *Orbis*, Fall, 1991.

May, Lynwood, "New Directions for the People's Republic of China Space Program," *Signal*, December 1987.

McLean, Alasdair, "A New Era? Military Space Policy Enters the Mainstream," *Space Policy*, 2000.

"Megawatt Laser Test Brings Space-based Lasers One Step Closer," *Space Daily*, April 26, 2000, http://www.spacedaily.com/news/laser-00e.html.

Meyer, S. M., "Soviet Military Programs and the "New High Ground," *Survival*, Volume 25 Number 5 (September/October 1983).

———. "Soviet Strategic Programs and the U.S. SDI," *Survival*, November/December 1985.

Ming, Chenug Tai, "Decimated Ranks: Peking Further Cuts PLA in Bid to Modernise Military," *Far Eastern Economic Review*, February 27, 1992.

Mohr, Charles, "'Option' Sought to Deploy Space Shield Soon," *New York Times*, October 19, 1986.

Moore, George, C. Budura, J. Johnson-Freese, "Joint Space Doctrine: Catapulting into the Future," *Joint Force Quarterly*, Summer 1994.

Moore, Thomas, "The Missile Defense Act of 1995: The Senate's Historic Opportunity," *The Heritage Foundation*, August 1, 1995.

"Navy Shift Elevates Space," *Aviation Week and Space Technology*, April 8, 2002.

Nguyen, Hing P., "Russia's Continuing Work in Space Forces," *Orbis*, Summer 1994.

Ning, Li, "30 Years of Development in Space Technology," *Beijing Review*, June 19, 2000.

Oberg, James, "The Heavens At War," *New Scientist*, June 2, 2001.

Oberdorger, Don, "Top-Level Fight Led to ABM Policy Shift," *Washington Post*, October 17, 1985.

Opall, Barbara, "PLA Pursues Acupuncture Warfare," *Defense News*, March 1, 1999.

Owens, Mackubin T., "Technology, the RMA and Future War," *Strategic Review*, Spring, 1998.

Owens, William A., "The Emerging U.S. System of Systems," National Defense University, *Strategic Forum*, Number 63, Institute for National Strategic Studies, February 1996.

———. "Revolution in Military Affairs? U.S. Vision for Future Warfare," *Revolution in Military Affairs? Challenges to Governments and Industry in the Information Age*, Conference Paper, Royal Institute for International Affairs, May 21-22 1997.

Payne, Keith B., L. Vlahos and W. Stanley, "Yeltsin's Global Shield: Russia Recasts the SDI Debate," *Policy Review*, Number 62, Fall 1992.

Perry, G. E., "Russian Hunter-Killer Satellite Experiments," *Royal Air Force Quarterly*, Winter 1977

Pike, Gordon, "Chinese Launch Services: A User's Guide," *Space Policy*, May 1991.

"Plasmadynamics and Lasers," *Aerospace America*, December 1999.

Rao, Radhakrishna, "China's Space Plan," *Satellite Communications*, February 1987.

Rathmell, Andrew, "Mind Warriors at the Ready," *The World Today*, The Royal Institute of International Affairs, November 1998.

Richardson, Robert C., "High Frontier: "The Only Game in Town"," *Journal of Social, Political and Economic Studies*, 1982.

Richter, Paul, "China May Seek Satellite Laser, Pentagon Warns," *Los Angeles Times*, November 28, 1998.

Rip, Michael R., "How Navstar Became Indispensable," *Air Force*, November 1993.

Roos, John G., "Militarizing Space," *Armed Forces Journal International*, September 2001.

Roy, Kenneth, "Ship Killers from Low Earth Orbit," *U.S. Navail Institute Proceedings*, October 1997.

"Russia Turns Military Ambitions to Space," *STRATFOR*, May 21, 2001, www.stratfor.com.

"Russia opens space "window" in Tajikistan," *SpaceDaily*, July 19, 2002, www.spacedaily.com/news/020719080734.rk4cuey0.html.

"Russians Alter MiG-31 for ASAT Carrier Role," *Aviation Week and Space Technology*, August 17, 1992.

"Russian ASAT," *Washington Times*, June 18, 1999.

Santoli, Al, "PLA Successfully Tests Advanced Antisatellite Weapon," *China Reform Monitor,* American Foreign Policy Council, Washington, D.C., No. 355, January 17, 2001 [http://www.afprc.org/crm/crm355.htm].

―――. *China Reform Monitor*, American Foreign Policy Council, Washington, D.C., No. 383, May 14, 2001 [accessed] http://www.afprc.org/crm//crm 383.htm.

"Satellite Killers," *Aviation Week and Space Technology*, Volume 104 Number 25, June 21, 1976.

Schweizer, Peter, "The Soviet Military Goes High-Tech," *Orbis*, Spring 1991.

Schroeer, Dietrich, "Directed-Energy Weapons and Strategic Defense: A Primer," *Adelphi Papers 221*, International Institute for Strategic Studies, Summer, 1987.

Scott, William B., "Space Warfare Center Aims to Be "Nellis of Space," *Aviation Week and Space Technology*, September 1, 1997.

―――. "USAF Space Battlelab Assessing New Concepts," *Aviation Week and Space Technology*, September 1, 1997.

―――. "New Satellites Sensors Will Detect RF, Laser Attacks," *Aviation Week and Space Technology*, August 2, 1999.

―――. "Cincspace: Focus More on Space Control," *Aviation Week and Space Technology*, November 13, 2000.

―――. "Wargames Zero In On Knotty Milspace Issues," *Aviation Week and Space Technology*, January 29, 2001.

―――. "USAF Warned to Bolster or Lose "Space Force" Franchise," *Aviation Week and Space Technology*, January 29, 2001.

———. "Wargame: "Space" Can Deter, Defuse Crisis," *Aviation Week and Space Technology*, February 5, 2001.

———. "Milspace Comes of Age in Fighting Terror," *Aviation Week and Space Technology*, April 8, 2002.

———. "Milstar Ring to Speed Data Toward Combat Zones," *Aviation Week and Space Technology*, January 21, 2002.

———. "Space Enhances War on Terrorists," *Aviation Week and Space Technology*, January 21, 2002.

"Service More Tight-Lipped About MIRACL's ASAT Mission," *Inside The Army*, Volume 10 Number 47, November 30, 1998.

Sheldon, Charles S., "Soviet Military Space Activities" *Soviet Space Programs 1971-1975*, Committee on Aeronautical and Space Science, U.S. Senate 1976.

Shulin, Deng, "The Development of China's Space Industry," *China Today*, November 1994.

Sietzen, Frank,"Laser Hits Orbiting Satellite in Beam Test," *Space Daily*, October 20, 1997 http://www.spacedaily.com/news/laser-97a.html.

———. "Lack of Space Assets Limits Chinese Military," *China In Space*, August 10, 1999, http://www.space.com/news/china_810.html.

Smart, Jacob E., General Ret. USAF, "Strategic Implications of Space Activities," *Strategic Review*, Fall 1974.

Sofaer, Abraham, "The ABM Treaty and the Strategic Defense Initiative," *Harvard Law Review*, June 1986.

"Space for Hire," *The Times*, (London: U.K.), April 14, 1994.

"Space-Based Laser Team Advances Design With Successful Test," *Space Daily*, January 25, 2001, http://www.spacedaily.com/news/laser-01a.html.

"Space-based Laser (SBL)," *Federation of American Scientists*, http://www.fas.org/spp/starwars/program/sbl.html.

Spencer, Jack, "What Foreign Leaders Are Saying About Missile Defense," *The Heritage Foundation*, July 12, 1999.

Spring, Baker, "For Strategic Defense: A New Strategy for the New Global Situation," *The Heritage Foundation*, April 18, 1991.

Taverna, Michael A., "Russian Aerospace Wants to Come in from Cold," *Aviation Week and Space Technology*, August 23, 1999, http://www.awstonline.com.

———. "Pacts With China, Italy Spotlight Latin American Space Ambitions," *Aviation Week and Space Technology*, October 9, 2000.

———. "India, China to Expand Earth-Observing Nets," *Aviation Week and Space Technology*, October 29, 2001.

Tomes, Robert R., "Revolution in Military Affairs—A History," *Military Review*, September/October 2000.
"TRW Wins Space Laser Contract," *Space Daily*, March 19, 1998 http://www.spacedaily.com/new/laser-98a.html.
"TRW Conducts Tests to Validate Laser Technology," *Aviation Week and Space Technology*, May 1, 2000.
Tirpak, John A., "Brilliant Weapons," *Air Force*, February 1998.
Tsipis, Kosta, "Laser Weapons," *Scientific American*, December 1981
Van Cleave, William R., et al, "C.I.S. and Nuclear Weapons: Liabilities, Risks, Proliferation and Strategic Defense," International Security Council Conference, *Global Affairs*, Volume 8 Number 1, Winter 1993.
Vizard, Frank, "Return to Star Wars," *Popular Science*, April 1999.
Wall, Robert, "Army Considers Lasing Satellite," *Aviation Week and Space Technology*, December 20, 1999.
———. "U.S. Laser Weapons Industry is Shrinking," *Aviation Week and Space Technology*, April 10, 2000.
———. "Space Reformers Juggle War, Acquisition Demands," *Aviation Week and Space Technology*, April 8, 2002.
———. "Space-Based Interceptor Gets New Lease on Life," *Aviation Week and Space Technology*, August 14, 2001.
———. "Killing Missiles at the Speed of Light," *Aviation Week and Space Technology*, August 14, 2001.
Weber, Steve, "ASAT Proponents Fail to Reverse White House Policy," *Space News*, September 19-25, 1994.
Weiping, Shi, "Achievements of Defense Conversion in China's Aerospace Industry," *Aerospace China* (English edition), Summer 2001.
Will, George F., "The Arms Control Fetish," *Washington Post*, September 21, 1995.
Wilson, J. R., "Putting Space Weapons on the Fast Track," *Aerospace America*, July 2001.
Wright, Jonathan, "U.S. Military Moves Closer to Star Wars," *Daily Telegraph*, May 21, 1998.
Xinzhai, Zhang, "The Achievements and the Future of the Development of China's Space Technology," *Aerospace China*, Summer 1996.
Xuan, Wu, "China's Space Technology Going Forward," *Aerospace China*, (English edition), Autumn 2001.
"Yeltsin Signs Lease on Baikonur," *The Financial Times*, London: U.K., March 29, 1994.
Yilin, Zhu and Xu Fuxiang, "Status and Prospects of China's Space Program," *Space Policy*, Februrary 1997.
Yilin, Zhu, "Fast Track Development of Space Technology in China," *Space Policy*, May 1996.

———. "The Development of Chinese Satellites under Professor Tsien," *Journal of The British Interplanetary Society*, volume 50, 1997.

Zhiqiang, Gao, "China Space Development Strategy," *Aerospace China*, (English edition) Winter 2000.

Index

ABM Treaty interpretations, 45-51
ABM Treaty withdrawal, 196
Afghanistan and military space, 206-205
ABM Treaty, 44-51
antisatellites, 15, 16; antisatellite and strategic stability, 110; Eisenhower and antisatellites, 110-111; Kennedy and antisatellites, 112; Ford and antisatellites, 114; Carter and antisatellites, 115; Reagan and antisatellites, 113-117; F-15 antisatellite, 116-118; U.S. antisatellites during the Cold War, 110-117; Soviet antisatellites during the Cold War, 117-125; antisatellites and arms control, 125-129; U.S. post-Cold War antisatellites, 130-146; kinetic energy antisatellites, 130-131; Mid-Infra-red Advanced Chemical Laser, 131-132; Airborne Laser and antisatellites, 133; Russia and antisatellites post-Cold War, 134; China's antisatellite program, 101-102
Arms Control in Space, 196-197

Brilliant Eyes, 22

Brilliant Pebbles, 22, 199-200

China's communications satellites, 88
China's intelligence satellites, 87
China's launch sites, 91-92
China's launch vehicles, 89-90
China's space cooperation, 96-101
China's space forces, 92-95
China's space organization, 92
Clinton and Missile Defense, 186-191
Clinton's Military Space Policy, 191-195
Commission to Assess U.S. National Security Space Management and Organization (Rumsfeld Commission), 197-199
Corona, 13

Defend America Act of 1996, 186
Defense Against Antisatellites (DSATS), 221-224

Exoatmospheric Kill Vehicle, 189-190

Fletcher Panel, 19-20
Fractional orbital bombardment system, 58

250 Index

G. W. Bush and Missile Defense, 195-196
Global Protections Against Limited Strikes, 23-25, 27-29
Gulf War and military space, 165-169

high ground, 13
Hoffman Panel, 19-20

Long March launchers, 86, 89-90

Midas, 13
Missile Defense "3 plus 3" program, 187-189
Missile Defense Act of 1991, 24
Moscow ABM site, 41

National Missile Defense Act of 1995, 186
National Missile Defense Act of 1999, 190-191
National Missile Defense, 189
NIKE-ZEUS, 36-37
Nuclear Weapons Effects in Space, 223-224

Outer Space Treaty, 15, 16

Revolution in Military Affairs, 170; definition of 170, and United States, 169-171; elements of, 171-172; and Russia, 177-178; and China, 176-180
Russia's space assets, 74-77
Russian military space, 68-72

SAFEGUARD, 39
SALT I, 16
Samos, 13

satellite reconnaissance, 13
SENTINEL, 38
Soviet Ballistic Missile Defense, 40-44
Soviet BMD radars, 42, 43
Soviet cosmodromes, 66-67
Soviet military space, 44-68
Soviet space control, 57
space control, 12
Space policy, 13-18, Eisenhower space policy, 13, 14; Kennedy space policy, 15; Johnson space policy, 16; Nixon space policy, 16; Ford space policy, 16; Carter space policy, 17; Reagan space policy 17-18
space power theory, 12-13
Space Reorganization, 194-196
space sanctuary, 12
space survivability, 12
space-based defenses, 25-27
Space-based laser 140-160, 199; science and technology of, 141-143; lethality of, 143-158; components of, 147-150, and industry, 151-153; development of, 154-160; and Missile Defense Agency, 160
Space-based missile defenses, 199-200
Space-based weapons against terrestrial targets, 200-202
SPARTAN, 38
Sputnik, 14
Strategic Defense Initiative Phase I, 18, 20, 21, 22

Strategic Defense Initiative, 18

Tsien Hsue-shen, 84-85

U.S. Ballistic Missile Defense, 36-40
U.S. Long Range Plan, 200
U.S.-Russian Cooperation in Space, 72-74

White, General, 14

Yeltsin's Global Protection System, 60-64
Yugoslavia campaign and military space, 202-204

About the Author

Matthew Mowthorpe is currently working for the Ministry of Defense in the United Kingdom. He has contributed to the OECD project on The Commercialization of Space and the Development of Space Infrastructure: The Role of Public and Private Actors. He has recently completed a Ph.D. at the Center for Security Studies, University of Hull on the Militarization and Weaponization of Space. Prior to this, he completed a MSC at the Department of Defense and Strategic Studies at Southwest Missouri State University. He has published articles on military space issues in the following journals: *Aerospace Power Chronicles, Space Policy, Journal of Social, Political, and Economic Studies, Journal of Slavic Military Studies,* and *National Defense University Journal of the Republic of China.*